Recent Advances in Oilfield Chemistry

Recent Advances in Oilfield Chemistry

Edited by
P. H. Ogden
Akzo Nobel Chemicals Ltd

THE ROYAL
SOCIETY OF
CHEMISTRY

Sep/ae
Chem

The Proceedings of the Fifth International Symposium on Chemistry in
the Oil Industry, organised by the Industrial Division of the Royal
Society of Chemistry, held in Ambleside, Cumbria on 13–15 April 1994

Special Publication No. 159

ISBN 0-85186-941-6

A catalogue record for this book is available from the British Library

Published by the Royal Society of Chemistry,
Thomas Graham House, The Science Park, Milton Road,
Cambridge CB4 4WF, UK

Printed in Great Britain by Bookcraft (Bath) Ltd

Preface

The market for oilfield chemicals has been estimated at approximately 100 million pounds sterling per year in the North Sea sector and ten times this value elsewhere.

This market consists of various chemical types used in several different applications. Since several of these chemicals are commodities such as cement, barite or salt, it is clear that the volume of individual speciality chemicals is rather small when compared with other industrial applications such as agriculture, detergents or pharmaceuticals. Consequently, in many instances the projected return on research investment is too small to justify longer term research into the development of industry specific materials and investigations are confined to ECOIN registered materials. This, together with tightening environmental constraints which prohibit discharge of many hitherto popular materials into our oilfield locations makes this industry frustrating for enthusiastic research chemists.

However, the exploitation of oil reserves, particularly those in hostile environments, is increasingly dependent upon the combined efforts of chemists and petroleum engineers, which has demanded the development of closer co-operation between major oil producers and service companies. At this symposium chemists and engineers presented up to date reports of several pertinent developments.

It was not possible to cover all aspects of the oil winning industry in the time available and we have concentrated on those areas where environmental pressures seem to be greatest.

I am indebted to fellow members of the organising committee:
Mike Fielder (BP, UK); Paul Gilbert (Shell Chemicals); Ian Macefield (Allied Colloids); Reg Minton (BP, Norge); and Terje Schmidt (Statoil) for their help and advice.

Paul H. Ogden
Akzo Nobel Chemicals Ltd
Littleborough
Lancashire

Contents

New water clarifiers for treating produced water on North Sea   311
production platforms
D.K. Durham

Subsurface disposal of a wide variety of mutually incompatible   317
gas-field waters
G. Fowler

Subject Index   329

A Review of Developments in Oilfield Drilling

R. C. Minton

BP NORGE UA, STAVANGER, NORWAY

Introduction:

The upstream sector of the Oil and Gas business is experiencing a period of rapid change. Cost pressures are intense and environmental issues are becoming more dominant. These conditions are forcing the Operating companies to re-evaluate their present business practices and, in several cases, radical changes to the former status quo are being made. This is having a knock on effect on the Service companies which will, in turn, affect the other companies in the supply chain.

At the business level the former concepts of chemical product supply, at a tendered price, are breaking down. In the drilling fluids area, individual chemical prices are no longer detailed. The income of a drilling fluids company is more and more related to a finished barrel or cubic meter of 'mud' and, under the more radical proposals now being evaluated, will be directly related to the 'footage' drilled. Consequently their financial reward will be linked to performance; not to the number of 'sacks' of chemicals they can utilise. These changes will have a subsequent bearing on the number, and specification, of the products that the Drilling fluid service companies market, and , subsequently, impact the chemical manufacturing and specialist supply companies.

Environmental considerations are similarly having a major impact on the drilling fluids business. Attention is focused on the occupational hygiene aspects of the fluids in use and on the nature of the wastes generated. Onshore, disposal of the generated cuttings and the waste drilling fluids is a major consideration when fluids are selected for use and a similar situation exists offshore. In the latter case, cuttings and spent drilling fluids can be discharged overboard under specified conditions. However, drilling fluid chemicals can not be selected solely on the basis of environmental restrictions, since their performance in the drilling operation is paramount.

These two areas, changing business practices and environmental pressures, will radically change the nature of the drilling fluid supply business and future profitability of the service companies will depend on how they approach these issues.

Business Environment:

The value of a barrel of oil is now lower in money of the day terms than at any time since 1979 (Fig.1) and, in real terms, it is lower than it was in 1974, immediately after the first 'Oil shock'. For the UKCS, as in all other areas, this has resulted in a significant reduction in the value of the sales from the existing producing fields (Fig.2), and is having a major impact on all aspects of the upstream business.

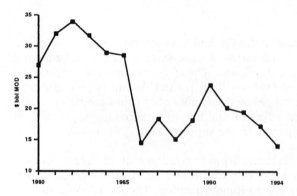

Fig.1 Average Crude Oil Prices (Brent) - 1980/'94

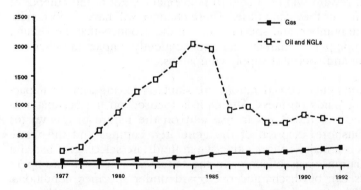

Fig.2 The Value of UKCS Oil & Gas Sales - 1977/'92 (Ref.1)

There were fewer wells drilled in the US in 1992 than at any time since 1966 (Fig.3) and the footage was the least since 1949. A limited increase was recorded in 1993 but this was still less than a third of the peak activity of 1981. Predictions for 1994 are still being revised downwards. Closer to home, the Exploration and Appraisal activity on the UKCS in 1993 resulted in 52 wells being drilled, down from 90 in 1992 and the demand picture for mobile drilling units in 1994 continues to show a downward trend (Fig.4), with resultant softening of the day rates for the mobile drilling units. There is no evidence of an aggressive E&A programme of work from any of the oil and gas companies and, consequently, this position is unlikely to change in the near term.

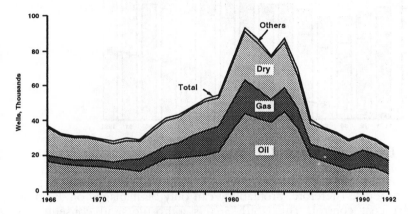

Fig.3 U.S. Drilling since 1966 (Ref.2

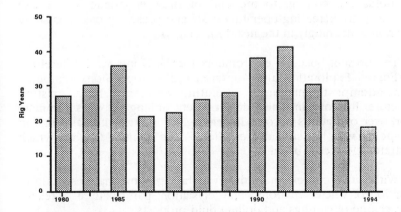

Fig.4 U.K.C.S. Semi-submersible Demand - 1983 / '94 (Ref. 3)

Field developments are also being scrutinised. Fields on the UKCS, that started production in the period between 1986 and '92, had a total cost per barrel of £13, almost $20 per barrel compared with today's market price of $14! New fields under development at the end of 1992 had a lower overall cost. However, at £8, or approximately $14, this is perilously close to today's market price.

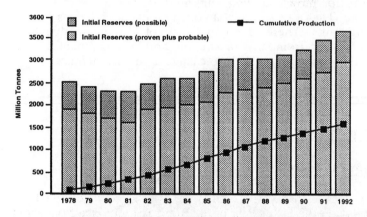

Fig.5 Cumulative U.K.C.S. Production versus Reserves

As an industry we have little, or no, control over the market price of our product. Therefore, profitability can only be managed on the basis of strict cost control. The North Sea is a hostile, and consequently expensive, environment in which to operate and, as the area matures, there are fewer large oil and gas accumulations to exploit. We have already produced about half of the proven reserves of the North Sea (Fig.5) and several fields are coming to the end of their economic life. These situations lead to increasing operating costs and present projections show these climbing alarmingly in the near term (Fig.6).

The present business environment for the North Sea is. therefore, one of limited Exploration and Appraisal drilling operations combined with an attempt to maximise operating efficiency from existing infrastructure. It is one in which development drilling activities dominate and workover operations become increasingly important. It is also one in which spend will be critically evaluated and limited to that which demonstrates clear cost benefits.

Within this context the industry is facing additional costs due to the adoption of stricter environmental legislation, particularly as it relates to the discharge of cuttings and drilling fluid products.

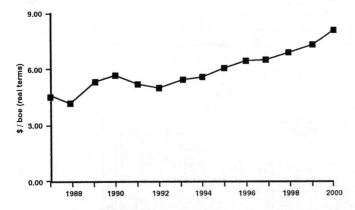

Fig.6 Projected U.K.C.S. Operating Costs (Ref.4)

Environmental Issues:

In the decade from 1983 to 1993 the average footage drilled per operating rig day, in the U.K. sector of the North Sea, doubled from about 60 feet to 130. Much of this is attributable to the widespread use of the low toxicity oil based muds that prevailed in both the exploration and development drilling operations. There were several aspects to the use of these fluids. Their primary benefit lay in their ability to control the swelling tendency of the Tertiary shales common to the North Sea. This characteristic reduced the time taken to drill these intervals, permitted longer open hole sections and allowed for slim hole well geometries. Benefits were also seen in the less reactive formations with faster rates of penetration and fewer lost time incidents. Figure 7 presents some typical data for a series of comparable wells in the Norwegian sector of the North sea drilled with inhibitive water based fluids and LTOBMs and clearly demonstrates the performance benefits of the latter. Similar data is available from a variety of sources.

Unfortunately, in using the LTOBM, and discharging the oil contaminated cuttings, there is a degree of localised environmental damage (Ref. 5) that is seen to be unacceptable. Controls on the discharge of these cuttings have therefore been imposed across the North Sea. It is no longer acceptable to discharge them at single well sites and there is a progressive programme to phase out discharges at multi-well development sites. In Norwegian waters these restrictions already apply whilst, on the U.K.C.S., full implementation will take effect in 1997.

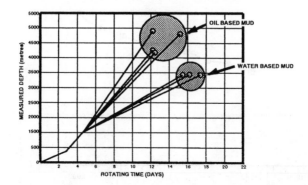

Fig.7 Drilling Times for Norwegian Appraisal Wells

The challenge faced by the industry is, therefore, to develop new approaches that avoid the discharge of contaminating products whilst maintaining, or improving, the historic drilling performance and minimising the addition spend.

Three approaches have been developed, each with their own advantages and disadvantages. The so called Pseudo-OBMs have been formulated such that they perform similarly to the mineral OBMs, without the same degree of environmental impact from the discharged cuttings. Secondly, novel water based drilling fluids have been developed, and field trialed, demonstrating improved performance relative to the earlier water based fluids. Both of these approaches will be discussed in greater detail later so they will not be addressed further here. The third approach, and one that is gaining greater acceptance across the industry, is that of cuttings slurrying and re-injection. This approach, depicted in figure 8, has been applied for the last two years off the Gyda platform in Norway and, in that period, no contaminated cuttings have been discharged (Ref.6)

This approach offers a final solution to the issue of overboard discharges and permits the continued use of the mineral OBMs. The logistics of the process make it an ideal solution for multi-well sites but it is less applicable for single well operations. For these, solvent extraction cleaning systems and thermal processes have been developed (Ref.7) such that the cuttings can be discharged without significant environmental impact. These processes are close to commercialisation. Another option, and one that is particularly attractive where the necessary infrastructure is available, is to take the cuttings from the single well operation to a platform where injection is taking place and dispose of them there. The cuttings can either be transported in a dry form or slurried and pumped from and to the different installations. Again, this new approach offers the opportunity of continued LTOBM use with the associated operational benefits.

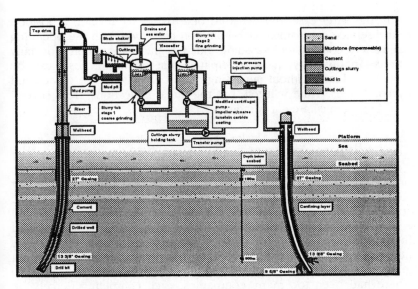

Fig.8 Cuttings Slurrying and Re-injection Schematic

To date the drilling fluid companies have not involved themselves significantly in this debate. They have focused primarily on the reduction in toxicity of the chemical species supplied and on the development of new drilling fluid formulations to meet the toxicological requirements that have been set. However, with the new approach to business that is developing there is greater need for them to be a part of the solution.

New Contracting Strategies:

This overall picture, which in many cases is as challenging as at any time since the early '70s has forced the operating companies to re-evaluate their business approaches. It is leading to new relationships between them and the service companies that support the business and will have ramifications for the whole of the service and supply side of our business.

The discussion that follows is specifically aimed at the drilling side of the upstream business. However, much of it is equally applicable to the other sectors in which the chemical suppliers and service companies act.

The response of the operating companies to this business environment has been varied, but two particular initiatives have been widely discussed. These are the 'Win '90s' approach, being adopted by Shell, and the 'Partnering' approach, being adopted by BP Exploration.

In the latter case, the declared objective is to develop a position where the service companies have common objectives with the operating companies and consequently assume much of the responsibility for daily operations. In this way, it is anticipated that operating costs can be substantially reduced without impacting the profitability of the service and supply companies. Time will tell, but the initial indications, from several world-wide locations, are encouraging.

Two years ago BP Alaska realised that it had to make major changes if the Prudhoe Bay field was to remain profitable beyond the year 2005. Not because the reserves would be depleted, but because of the inexorable rise in operating costs. An early exercise involved combining BP and ARCO's staff resources in the areas of procurement, warehousing, aviation, maintenance and drilling, into a single organisation. This saved some $50 - $60 million during 1992 and '93. However, this was still insufficient! Consequently, BP Alaska formed a number of strategic alliances with selected service companies so as to obtain optimum efficiencies for equipment, facilities and personnel in the area. Prior to this, there had been too many companies, each competing fiercely on price, and, consequently, inefficient use of resources. However, the downside of this is that a number of service companies in the area have now gone out of business! For BP there have been annual savings of 20% in the costs of well services and, for the remaining service companies, there have been savings associated with optimum resource use. They can now take a more strategic view of their business rather than having to rely on short term contracts. In one recently documented case (Ref.8) the annual costs associated with wireline operations in Alaska have been reduced by $2 million whilst the service company involved is enjoying a stable, predictable, profit.

The Miller field is another example of the new contracting approaches being adopted by BP, with a somewhat different emphasis to that of the Alaskan model. On Miller the drilling and drilling services have been awarded to Sante Fe Drilling and Baroid EDS. The contract is structured in such a way as to 'reward' improved performance and reduce profitability when the drilling performance is poor. Lump sum payments reward early well delivery and are balanced by penalty rates for well duration overruns (Ref.9). Within this remuneration envelope, the decisions as to how the well is to be drilled are taken by the consortium, in conjunction with BP. The point to note, however, is that the reward is strictly aligned to performance and not to consumption. Hence, the excessive use of products, that would formerly have added to a company's profits, now reduces their profitability! This has ramifications for the selection of products, in particular the drilling fluid, workover and production chemicals, as the market matures. Selection will be much more critically aligned to performance in future.

The approach adopted for Miller has resulted in a 45% reduction in drilling and completion costs and a 24% reduction in drilling times, relative to the offset template drilling activity (Fig.7), whilst providing 'good business' for the two service / supply companies.

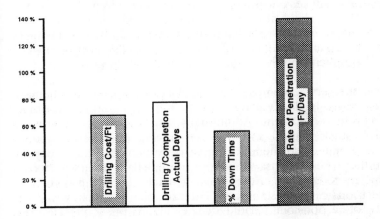

Fig.9 Comparison of Performance relative to Earlier Template Drilling Operations

A number of further examples could be quoted so as to define the changing relationships that are being established under these new contracting strategies. Similarly, other operators would be able to identify their own approaches to the challenge set by the poor business environment that presently prevails. The key message, however, is that the business drivers for the service companies are changing dramatically. For the drilling fluid service companies this means that their income relates directly to performance in terms of 'footage' drilled or at least to the volume of fluid used, irrespective of the subsequent addition of conditioning chemicals!

Future Drilling Fluid Service Areas:

As of today, most contracts between the drilling fluid service companies and the operators are based on defined unit prices for the individual chemicals, and a daily engineering charge. There is therefore no direct incentive to minimise chemical consumption nor to speed up the drilling process through innovative engineering. With the new contracts, however, the situation is completely reversed. Gross income for the service will be defined on the basis of a footage charge or, at least, on the basis of the volume of fluid used over a hole section. Profitability will, therefore, depend on minimising the consumption of chemicals and the engineering time required. This provides a direct alignment between the objectives of the service companies and the operating companies in respect of the service being delivered.

Business success will be more closely aligned with the performance of the individual chemicals and there will undoubtedly be a reduction in the number of speciality products used and the range of products stocked. This clearly has a bearing on the relationships between the chemical suppliers and the service companies.

Since the service company has an incentive to reduce chemical consumption, there is clearly a need for them to take the responsibility for solids control equipment. Additionally, areas such as mud pump consumables, stock control and the use of chemical dosing systems become of importance to them. The area of responsibility can then be widened further to encompass the whole supply chain from selection of chemical species to disposal of the wastes generated. In this way the true value of decisions can be established. For example, a company that has this breadth of responsibility could decide that an onshore facility for centrifuging the fluid and conditioning it chemically, with fluid being continually changed out at the rig, is more cost effective than offshore treatment. If this has no adverse effect on the drilling performance and offers them a better return for their business then there is no reason why this approach could not be adopted. This plant could then be used to support a number of operations. Under the 'old' contracting approach this option did not exist. Now the range of options is solely limited by our imagination.

The other major area of change will be in respect of data collection and analysis. It will now be critical that the companies have an accurate appreciation of the performance of different chemicals both in order to compare the efficacy of the products from different suppliers and to judge the benefit of changing to different chemistries. For example, the question will now be asked as to the relative merits of technical grades versus the purer products and, for liquid products, as to the minimum solvent levels that can be used so as to minimise handling, transport and stocking requirements.

These are all areas in which it would be natural for the drilling fluid service companies to accept, and indeed require, responsibility. However, there are synergies with the mud logging companies and with the cementing companies that could be explored to good effect. In each case the criteria would be: 1) Does the approach add value to the operation and 2) is it an attractive business proposition for the service company? If these conditions are met then the approach will be adopted.

This paper commenced with a fairly pessimistic view of the business environment facing the industry. However, the consequence of this position could lead to a far better business position for all concerned. There is an opportunity for those companies that have the technical and commercial skills to improve their level of profitability and widen their scope of responsibility whilst, at the same time, meeting the operator's needs for an efficient drilling operation. We all need to appreciate that the business has to change and that there will be an initial period of uncertainty and doubt. The end result, however, is worth working for.

Conclusions:

1) The business environment demands a change in the historical approach the industry has taken in respect of the service companies.

2) Significant reductions in operating costs are required but this can not be achieved through further erosion of profit margins.

3) Cost pressures from environmental protection measures have to be managed and the choices established in business terms.

4) Partnering approaches have been shown to offer significant benefits to all parties involved.

5) Alignment of objectives is the key.

6) The roles and responsibilities of the drilling fluid service companies, in common with the other service areas, are changing. This will have a knock on effect to the companies in their supply chain.

Acknowledgement:

The author is grateful to BP Norge for permission to publish this paper and for the efforts of all his colleagues across the company who are striving to make these new approaches work.

References:

1) Anon. , <u>Development of the Oil and Gas Resources of the United Kingdom</u>, Department of Trade and Industry, 1993.

2) Anon., US Drilling: Industry needs to think positive, <u>World Oil</u>, Feb 1994.

3) Anon,. Semi-submersible market forecast, <u>FT North Sea Rig Forecast</u>, March 2, 1994.

4) L. LeBlanc, Operating budget increasing at expense of capital outlays, <u>Offshore</u>, Feb.1994.

5) J.Addy et.al., Environmental effects of Oil-Based Mud Cuttings. SPE Paper 11890, presented at Offshore Europe, Aberdeen, 1983.

6) R.C.Minton, Cuttings Slurrying & Re-injection - Two years of experience from the 'Gyda Platform'. Paper presented at the 7th Annual two day conference on Offshore Drilling Conference, November, 1993.

7) R.C.Minton, Technology Developments and Operational practices to minimise the Environmental Impact of Drilling Operations. <u>Proc.Royal Soc.of Chem., 150th Anniversary Ann.Chem.Congress,</u> London, 1991

8) C.Philips et.al., Strategic Alliances in the Wireline Services Industry, Paper SPE 27460 presented at the 1994 SPE/IADC conference, Dallas, Texas.

9) D.G.Nims et.al., BP's Well Construction Strategy - A Way Forward? Paper SPE/IADC presented at the 1994 SPE/IADC Drilling Conference, Dallas, Texas.

Mechanisms and Solutions for Chemical Inhibition of Shale Swelling and Failure

L. Bailey, P. I. Reid, and J. D. Sherwood

SCHLUMBERGER CAMBRIDGE RESEARCH, PO BOX 153, CAMBRIDGE
CB3 0HG, UK

1 INTRODUCTION

Shales are the most common rock types encountered while drilling for oil and gas and give rise to more drilling problems per metre drilled than any other type of formation. Estimates of the world-wide non-productive costs associated with shale problems are put at between $500 and $600 million dollars annually[1]. Common drilling problems include:

    hole closure
    hole collapse
    hole erosion
    bit balling
    poor mud condition.

In addition to the above drilling problems, the poor wellbore quality often encountered in shales may make logging and casing operations difficult.

Over many years the oil industry has responded by devoting large amounts of manpower and money to the study of the mechanisms of mud/shale reactions and to the development of solutions to control problems of shale instability.

Early studies concentrated on improving the performance of water-based muds (WBM) and culminated in the successful application of potassium chloride/partially hydrolysed polyacrylamide (KCl/PHPA) fluids in the late 1960s. These KCl/Polymer muds reduced the frequency and severity of shale instability problems to the extent that deviated wells in highly reactive formations could be drilled, although often still at a high cost and with considerable difficulty.

In the 1970's the industry turned increasingly towards oil-based muds (OBM) as a means of controlling reactive shale. Although (as with most new technology) there was a learning phase during which chemical instability still caused occasional problems, the formulation and engineering of these fluids rapidly became refined to the point where chemically active shales could be controlled through the use of appropriate surfactants and the correct salinity of the emulsified aqueous phase. The improved drilling performance obtained with OBM is well documented[2-4] and, although the cost benefit of using these fluids varies with geographical area and well design, it is usual for OBM to be viewed as the most economic option for all but the simplest of wells. OBM provides not only excellent wellbore stability but also good lubricity, temperature stability, a reduced risk of differential sticking and low formation damage potential, making it an essential tool for the economic development of many oil and gas reserves.

The use of OBM would have continued to expand through the late 1980s and into the 1990s were it not for the realisation that the discharges of oily wastes can have a lasting environmental impact. In many areas this awareness has led to legislation which prohibits or limits the discharge of these wastes and this, in turn, has stimulated intensive R&D activity to find environmentally acceptable alternatives. Three possible options have been considered: i) cleaning oily cuttings prior to disposal, ii) grinding and re-injecting oily cuttings at the rig site and iii) developing fluids with performance to rival that of OBM while still satisfying discharge requirements. The first two options allow the continued use of OBM and focus on the treatment of the waste products, whereas the last option — which includes improved WBM and muds with synthetic base fluids — deals with the problem at source by removing the need for OBM.

This paper is concerned with the development of improved WBM and focuses on the issue of shale inhibition, widely held to be the most critical shortcoming of conventional WBM. It is our view that the successful identification and application of highly inhibitive WBM would provide a cost-effective alternative to OBM in all but the most challenging of wells. Our aim is to gain a useful mechanistic understanding of the reactions which occur between complex, often poorly characterised mud systems and equally complex, highly variable shale formations. This paper discusses experimental and modelling studies in progress at Schlumberger Cambridge Research, and reviews other published results. We consider three processes contributing to shale instability:

  i) Movement of fluid between the wellbore and shale. Our discussion will
     be limited to flow *from* the wellbore *into* the shale
 ii) Changes in stress (and strain) which occur during shale/filtrate interaction.
iii) Softening and erosion, caused by invasion of mud filtrate and consequent
     chemical changes in the shale.

We review likely mechanisms for each process. Practical approaches to the selection, design and engineering of inhibitive fluids are discussed, and areas which would benefit from further innovation are identified.

2 INVASION OF MUD FILTRATE

The movement of fluid from the wellbore into the surrounding shale is controlled by differences between the chemical potentials μ_i of the various species within the wellbore, and the corresponding chemical potentials within the formation. The chemical potential μ_i can be written as

$$\mu_i = pV_i + RT \ln a_i \qquad (1)$$

where p is the thermodynamic pressure, V_i the partial molar volume of the ith chemical species, R the gas constant, T the temperature and a_i the activity of the species. The flux $f_i = \sum_j L_{ij} \nabla \mu_j$ of the ith species will depend on the gradients of all the chemical potentials and on a set of transport coefficients L_{ij}. Thus the fluxes depend upon gradients of both pressure and chemical activities. Transport into (or out of) the shale can be eliminated if the chemical potentials of *all* species in the wellbore fluid and in the shale are in equilibrium[5], and it is clear that transport can be reduced by modifying either

(a) the hydrostatic pressure within the wellbore, or
(b) the chemical composition of the wellbore fluid.

We need to know the relative importance of pressure and chemical composition, and how this depends upon the type of mud and shale. Various workers have emphasied the importance of either pressure, or composition, or a combination of the two. If the formation is chemically inert (e.g. sandstone), then invasion is controlled solely by the difference between the wellbore pressure and the pore pressure within the rock. Bol[1] presents evidence to support the dominant role of mud pressure in WBM invasion of shale. On the other hand, if the shale acts as a perfect ion exclusion membrane, only water can enter. In this limit, invasion is controlled solely by the difference between the chemical potential of water in the wellbore and within the rock, as suggested by Chenevert's early work[6]. Mody and Hale[7] take an intermediate view in which pressure has a major influence in poorly consolidated (more permeable) shales; water activity becomes increasingly important as the shale becomes more consolidated. Such compacted clays have a lower permeability, and are also likely to be more efficient at excluding ions[8]. To accommodate these changing responses, Mody and Hale made use of a reflection coefficient to quantify the extent to which the shale acts as an imperfect semi-permeable membrane. Knowledge of the likely existence of and, if it exists, the quality of the membrane is of critical importance when designing mud systems or developing new additives intended to restrict or eliminate fluid ingress. An understanding of the fluid invasion process has been the focus of much of our attention at Schlumberger Cambridge Research, and we now discuss two of our experimental techniques.

2.1 Wellbore simulation studies

A number of groups[9-12] have constructed equipment to study shale/drilling fluid interactions under conditions which approximate those downhole. SCR studies are carried out using our two smaller wellbore simulators: the SWBS[13,14] which uses a mechanical caliper to detect changes in the wellbore dimensions, and the X-ray transparent XWBS[15], which can be used at a CT facility to image invasion of fluid into the shale. The simulators have a common design, shown in figure 1. The shale sample (150 mm diameter, 200 mm long with a 25 mm central wellbore) is contained in a high pressure vessel. Test fluids are introduced to the central wellbore through a high pressure flow loop; flow rates of up to 15 l/min are possible. The SWBS maximum pressure is 31.5 MPa; the aluminium alloy pressure vessel of the XWBS operates up to 21.5 MPa. In the SWBS typical axial, confining and wellbore pressures are 27.3 MPa, 26 MPa and 25 MPa, respectively.

For our studies of shale/fluid interactions we usually use Pierre I. This is a reasonably strong and reproducible shale, and can provide the relatively large samples used in the

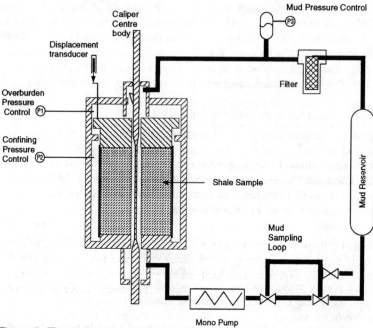

Figure 1. The wellbore simulator.

Age	Pierre I Cret.	Pierre II Cret.	London Tertiary	Oxford Jurassic
Mineralogy %				
Quartz	32	32	32	17
K-feldspar			2	7
Pyrite		2	1	5
Gypsum				1
Smectite	10	35	23	
Illite-smectite		9		17
Illite	30	16	29	30
Kaolinite	11	2	11	18
Chlorite				7
Moisture (%)	12	24	14	22

Table 1. Mineralogies of standard test shales.

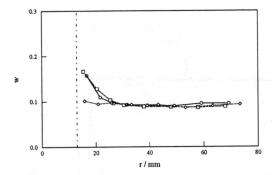

Figure 2. Measured final values of water content w as a function of radial position r, after a 48 h SWBS test in (i) ——O—— deionised water (ii) — — ☐ — — 5% KCl solution, (iii) ·····◇·····24% KCl solution. The broken vertical line indicates the initial position of the wellbore wall. From Sherwood & Bailey[14].

simulators. The mineralogy is given in table 1. As the shale samples are of outcrop material, they show little water sensitivity in the native state, and have a water activity $a_w = 1$. At the start of the experiment the shale sample is drained to reduce the water activity and increase the reactivity of the sample. Typically drainage under an isotropic pressure of 15 MPa over 5 days reduces a_w to 0.9.

The flow loop is now filled with drilling fluid, which circulates through the wellbore and can invade the shale. Changes in wellbore dimension during the test are monitored by a caliper, which also acts as a dummy drillstring, creating realistic flow conditions. A SWBS test typically lasts 48 hours. Post-mortem analysis of cores after a series of SWBS tests showed that invasion by water could be controlled by modifying the water activity of the wellbore fluid[14]. Figure 2 shows the extent of water invasion from simple polymer muds containing 0% KCl, 5% KCl and 24% KCl. The water activity of the 24% KCl solution equalled that of the shale as measured by a humidity meter. Hence, if the shale behaved as a perfect ion exclusion membrane, we would expect invasion of water from the 24% KCl solution to be absent, as is indeed observed. However, the analysis of cations within the shale (figure 3) shows that extensive invasion of potassium from the wellbore fluid has occurred with both the 4% and 24% salt solutions. Thus the shale is far from being a perfect ion exclusion membrane.

In another set of experiments both the water activity of the mud and shale were fixed (but not matched) and the pressure of mud in the wellbore varied. The pressure within a porous frit set in the shale approximately 5 cm from the wellbore wall was monitored. This work is still in progress, and is difficult because of the natural variability of the core. Results to date show that the measured pressure always attains the wellbore pressure, though the rate at which this is achieved depends upon the choice of wellbore pressure. This equilibration of pressure supports the findings of Ballard[12] who studied water movement in the higher permeability Oxford Clay at lower confining pressures, and concluded that

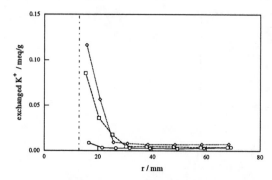

Figure 3. Measured final values of K^+ concentration as as a function of radial position r. Symbols as in figure 2.

osmotic processes played no part in the control of water ingress. Work is also underway to examine changes in the pressure/time relationship as a function of the water activity of the wellbore fluid.

2.2 Small-scale stress/strain tests

We have also studied the behaviour of shale under ambient temperature and pressure in a bench-top experiment. The device (figure 4) uses small cylindrical cores and is capable of measuring stress or strain changes when either preserved shale or reconstituted material is exposed to drilling mud. A full description of the device will appear elsewhere. The results presented here were obtained by measuring the unconfined linear swelling of preserved cores of Pierre I Shale with water activity $a_w = 0.90$, prepared in a constant humidity chamber.

Samples of shale were exposed to KCl solutions with activities of 0.95 and 0.85 and the swelling response observed. If swelling were controlled solely by the chemical potential of water, we would expect the shale to swell in the higher activity (0.95) KCl solution but to shrink in the fluid with lower activity (as pore fluid is drawn out of the shale). In fact, significant expansion of the shale was observed in both fluids, though the swelling was less in the more highly concentrated brine. These results are not fully consistent with those from the SWBS, but it should be noted that the preparation of the shale differed in the two experiments, and that the bench-top test is performed at ambient pressure. We again conclude that the shale does not behave as a perfect ion exclusion membrane.

Additional experiments were carried out with two oil based muds, in which the water activity of the internal brine phase was either $a_w = 0.95$ or 0.85. When cores with a water activity of 0.9 were exposed to these fluids, we observed swelling with the first fluid ($a_w = 0.95$) and shrinkage with the second ($a_w = 0.85$), as shown in figure 5. This behaviour supports the existence of an osmotic barrier at, or near the surface of the shale, presumably formed by the adsorption of surfactants present in the OBM. Further experiments supported the role of surfactants in producing a membrane and indicated its

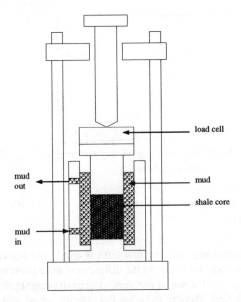

Figure 4. Schematic of the apparatus used for the small-scale stress/strain tests.

robustness: core with a water activity of 0.9 was first exposed to OBM with $a_w = 0.95$ and the expected swelling behaviour was observed. The OBM was then replaced by an aqueous calcium chloride brine with $a_w = 0.85$, which caused rapid shrinkage. This suggests that once the surfactant membrane is formed the shale can show perfect osmotic behaviour in WBM as well as OBM. When the experiments were performed using calcium chloride brines without pretreatment of the shale by OBM, swelling occurred in all cases.

3 THE EFFECTS OF FLUID INVASION

If invasion of mud filtrate occurs, we may examine the subsequent response of the rock to the invading fluid. There will be both physical and chemical effects, which we shall attempt to treat separately.

3.1 Stresses and strains

If a wellbore is drilled in a permeable, chemically inert sandstone, a 2-dimensional plane strain analysis gives the deviatoric stress at the wellbore:

$$\sigma_{rr} - \sigma_{\theta\theta} = -2(\sigma_{rr}^{\infty} + p_{\text{mud}}) + 2\eta(p_{\text{mud}} - p^{\infty}) \qquad (2)$$

where $\sigma_{rr}^{\infty} = \sigma_{\theta\theta}^{\infty}$ is the stress (assumed positive in tension) far from the wellbore, p_{mud} is the pressure of the drilling fluid within the wellbore, p^{∞} is the pore pressure at infinity, and η is a poroelastic parameter[16], with $0 \le \eta \le \frac{1}{2}$. At the wellbore the pore pressure is equal to p_{mud}, as is $-\sigma_{rr}$, so the left-hand side of (2) is $-(\sigma_{\theta\theta} + p_{\text{mud}})$ and is the effective

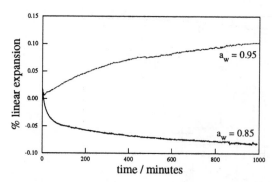

Figure 5. Small-scale strain test of Pierre I shale at $a_w = 0.9$ in OBM at (a) $a_w = 0.95$ and (b) $a_w = 0.85$.

tangential stress. On the right-hand side, the first term is due to the instantaneous elastic response of the rock, and the second is due to the diffusion of pore pressure into the rock. Hence, if $-\sigma_{rr}^{\infty} > p_{\mathrm{mud}} > p^{\infty}$, diffusion of pressure reduces the magnitude of $|\sigma_{rr} - \sigma_{\theta\theta}|$, but also raises the pore pressure, thereby reducing the effective stress and weakening the rock.

Bol[1] suggests that one reason why OBM gives greater shale stability than WBM is that the OBM does not wet, or enter, the shale pores and cannot therefore raise the pore pressure. If so, a differential pressure of the order of 4000 psi or more would be required to force oil into these very low permeability formations; such a value is clearly much higher than the overbalances experienced in drilling operations. However, this argument neglects the possibility that surfactants in OBM will cause the shale to become hydrophobic, thereby allowing fluid ingress and pressure increase. Invasion of OBM into shale is clearly seen in field samples such as cores and cuttings, suggesting that the shale stability seen in OBM is due more to the inert nature of the invading fluid than to the absence of penetration.

If the shale behaves as a perfect ion exclusion membrane, water flow will be driven by the difference between the chemical potential μ_w^{mud} in the mud, and μ_w^{∞} in the shale at infinity. It can be shown[14] that (2) becomes

$$
\begin{aligned}
\sigma_{rr} - \sigma_{\theta\theta} &= -2(\sigma_{rr}^{\infty} + p_{\mathrm{mud}}) + 2\eta(\mu_w^{\mathrm{mud}} - \mu_w^{\infty})/V_w \\
&= -2(\sigma_{rr}^{\infty} + p_{\mathrm{mud}}) + 2\eta[p_{\mathrm{mud}} - p^{\infty} + (RT/V_w)\ln(a_w^{\mathrm{mud}}/a_w^{\infty})]
\end{aligned}
$$

In general, ions (and other molecules) will invade the shale, and will affect the stress. These direct chemical effects are discussed in §3.2. The rock chemistry will also play an indirect role, by influencing the transport of the pore fluid. If the rock has an ion reflection coefficient $s \leq 1$, then two diffusion processes operate. The first is a diffusion of pressure, and the chemical contribution $(RT/V_w)\ln(a_w^{\mathrm{mud}}/a_w^{\infty})$ is multiplied by s, as predicted by Mody & Hale[7]. A subsequent, slower diffusion of solute then reduces the

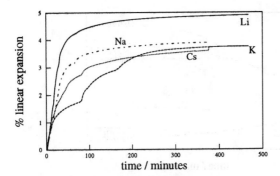

Figure 6. Small-scale strain test of Pierre I shale at $a_w = 0.9$ in 1 mol l^{-1} solutions of (a) LiCl, (b) CsCl, (c) KCl, (d) NaCl.

chemical contribution. Further details will be given elsewhere.

3.2 Changes in shale chemistry

A frequently used, though over-simplified indication of the effect of pore fluid composition on swelling can be based on electrical double layer theory: an increase in the salinity of the pore fluid will compress the double layers, and the shale will tend to shrink. A decrease in the salinity will increase the repulsions between clay particles. In this latter case more of the total externally applied stress will be borne via electrical stresses, and less at the points of contact (or cementation) between clay particles. The contacts therefore become weaker, reducing the strength of the rock in the same manner as a reduction in effective stress. However, it is now generally thought that the separation between clay particles in a compacted shale corresponds to the thickness of only a few layers of water molecules: electrical double layer theory is therefore inappropriate as it neglects clay hydration[17], and ion-specific effects[18]. Monte-Carlo[19] and Molecular Dynamics simulations[20] are beginning to give useful insights into the ordering of water between clay surfaces. Such studies suggest that the first few layers of water adjacent to the clay surface are highly structured, and responsible for the short-range repulsions between clay particles[17]. Different counterions have different hydration shells; changing the counterion will modify the water structure and the repulsive forces.

It is well known that cation type can greatly affect the swelling behaviour of clay minerals in water, and numerous studies have been reported[18,21]. In general, small, highly hydrated cations with a low hydration energy are more effective than those which are large and highly hydrated. The effectiveness of inhibition depends both on the extent to which the exchange form will swell and the selectivity of the clay mineral for that cation. Quirk[22] has shown that the swelling of various exchange forms in dilute suspension follows the order:

$$Li \sim Na > Ca \sim Mg > K > Cs.$$

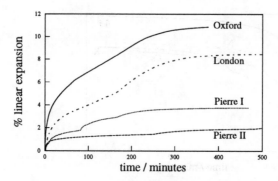

Figure 7. Small-scale strain tests of standard shales (a) Oxford, (b) Pierre I, (c) Pierre II and (d) London in 1 mol l^{-1} KCl solution.

In general, for a given cation oxidation state, selectivity follows the lyotropic series:

$$Li < Na < K < Rb < Cs \quad \text{and} \quad Mg < Mn < Ca < Sr < Ba.$$

In addition, it has been shown[23] that montmorillonite has a greater selectivity for K over both Ca and Mg ions.

The effects of salinity and salt composition on swelling stress and strain have been examined using core plugs of preserved Pierre Shale using the small-scale stress/strain tester described in Section 2 (figure 6). Note that at low salt concentrations the order of effectiveness of monovalent cations approximately follows the series given above with clear differences between lithium/sodium and potassium/caesium. At higher concentration the difference between the ions is much less pronounced, suggesting that any ion-specific behaviour at low concentration is swamped by general salinity effects when sufficient salt is present. For reasons of effectiveness, cost, availability, toxicity and compatibility with other mud additives, potassium salts are frequently used as swelling inhibitors in WBM[24]. The swelling response of several shales to potassium chloride solution is given in Figure 7 and the mineralogical composition of the shales in Table 1.

The exchangeable cations used to inhibit swelling may also include low molecular weight cationic polymers which are small enough to enter with the filtrate. These are remarkably effective in eliminating swelling, but suffer from rapid depletion due to their very high affinity for clay minerals.

Recently, organics such as glycerols and glycols have been shown to be highly efficient inhibitors, and are now being used extensively in WBM. The reasons for their effectiveness are not fully understood, and are the focus of much current research. Several mechanisms have been advanced, one of the more plausible being the ability of glycols to disrupt, through hydration bonding, the water structure which would otherwise build up adjacent to clay surfaces.

3.3 Softening and dispersion

Once the shale begins to react with the filtrate, the strength of the rock drops sharply as grain contacts are broken. If the rock fabric begins to fail, and repulsive forces drive the clay plates further apart, the shale becomes more permeable, thereby increasing the rate at which swelling can occur. This will continue until the clay is completely dispersed. At these large separations electrical double layer theory becomes applicable[25]. Moreover, once weakened, the shale is susceptible to mechanical or hydraulic forces arising from the circulating mud or from contact with the drillstring; such forces will mechanically disperse the shale.

The dispersion can be minimised by high ionic strength drilling fluids, which screen out the electrostatic repulsions between clay particles. Alternatively, high molecular weight polymers may be used. These adsorb on the clay surface and limit dispersion by a bridging mechanism. This is the major function of PHPA polymers, which have proved highly effective when used in conjunction with KCl.

An interesting alternative approach may be to use mud additives which undergo reactions with the clay minerals and/or pore fluids present in shales to produce cements which strengthen the rock to prevent failure. Silicate and phosphate salts have been evaluated for this application and have been demonstrated to have potential in field trials, although some drilling difficulties not related to wellbore stability have been reported[26]. Glycols have also been reported to harden shale, although the way in which this is achieved is as yet unclear.

4 CONSEQUENCES FOR MUD DESIGN

4.1 Mud design to minimise invasion

The shale studied in the tests discussed in §2 did not behave as a perfect ion exclusion membrane. The difference in water activity between the wellbore fluid and the pore fluid will therefore play a less important role (compared to the difference in pressure) than might be expected from the definition of chemical potential (1). This suggests that the major weapon presently available for controlling ingress of WBM filtrate into shale is mud weight, which should be kept as low as safety and mechanical wellbore stability considerations allow. Wherever possible, programmed mud weight increases through a shale section should consist of a series of small steps which keep the overbalance low at all times. This approach can give rise to a dilemma when cavings or tight hole problems are experienced: the mud weight needs to be raised to cure immediate problems but at the expense of increasing the invasion rate and therefore increasing the potential for more — and perhaps bigger — problems some hours or days later.

Conversely, with OBM, one should aim to balance the chemical potential of water in the wellbore with that in the shale, by choosing a suitable combination of mud weight (pressure) and mud chemistry (water activity).

If it is accepted that differential pressure is the critical factor governing the ingress of

WBM filtrate, it is easy to see how mud formulations could be developed which give better control:

(i) Creation of a semi-permeable membrane

If an effective membrane can be produced by adding suitable surfactants to WBM, then water ingress could be controlled using the activity of the filtrate, as in OBM. This effect was presumably obtained, at least to some degree, with the direct emulsion WBM which saw occasional field use up to the 1980s. The challenge is to identify effective surface active molecules which are environmentally acceptable, do not unduly affect other mud properties and, ideally, show low depletion rates.

(ii) Provision of fluid loss control

Conventional fluid loss polymers produce mud filter cakes which are typically one or two orders of magnitude higher in permeability than shales. If fractures are present, such polymers may be effective, but filter cakes are otherwise unlikely to form on shale. If they did, the shale — being the less permeable of the two solid phases — would still control the rate of fluid transport. This assumption is supported by experimental measurements[13,27] which show that common WBM polymers such as cellulose derivatives, Xanthan gum and PHPA do not slow water ingress appreciably. Given the small dimensions of pores in shales, it is likely that fluid loss control will best be achieved either by chemical reactions that greatly reduce — or even eliminate — permeability or by ultra fine particles small enough to block pore throats. Some suggestions along these lines can be found in recent oilfield publications[1,11,26].

(iii) Increasing the viscosity of the filtrate

A third approach would be to increase the viscosity of the filtrate (for example with sugars, silicates or glycols) such that the rate of ingress is reduced[1,27]. However, this serves only to slow rather than prevent fluid ingress and may not suffice to control wellbore stability.

4.2 Mud design to minimise subsequent swelling and dispersion

If invasion of a WBM filtrate cannot be avoided, the swelling response of the shale can be minimised by appropriate design of the filtrate chemistry:

(i) Control of ionic strength

The salinity of the filtrate should be at least as high as that of the pore fluid it is replacing.

(ii) Choice of inhibiting ion

Cations such as potassium (and caesium) should be incorporated into the formulation. These will replace the sodium, calcium and magnesium cations found in most shales to produce less hydrated clays with a significantly reduced swelling potential.

Although potassium ions reduce clay swelling, they rarely eliminate it completely. There is ample field evidence showing that a wide range of potassium concentrations fail to prevent chemical instability in reactive shales. Recent attempts to find more effective cations for the inhibition of swelling — e.g. aluminium complexes and low molecular weight cationic polymers — suggest there may be scope for further improvements, although it is not clear whether there are cost-effective solutions which are both environmentally acceptable and compatible with common mud additives.

Whatever inhibitors are added to the mud system, their concentration should be sufficiently high to remain effective as the filtrate travels through the shale. Depletion of an inhibiting ion (and its replacement by sodium, calcium and magnesium) as the front moves into the shale can be postulated as a cause of stress increase some distance into the rock, several days after the section is first exposed to the drilling fluid. This could be the cause of some so-called 7-day shales. Ion depletion and its link to stress changes within the body of the rock is a current area of investigation in our laboratory.

(iii) Structure breakers

If the concept of water structuring as a controlling clay swelling is accepted, the use of glycols and other small molecules capable of forming hydrogen bonds with water and clay surfaces is an important tool in inhibitive mud design. This area of chemistry will benefit from further research, and it is likely that by obtaining a good understanding of the mechanisms by which these materials operate, more effective drilling fluids and engineering practices will be developed.

(iv) Control of dispersion

The drilling fluid should incorporate an additive such as an encapsulating polymer, or a cementing agent, to minimise the disintegration of any weakened partially swollen shale. The current approach is to add a high molecular weight polymer, such as PHPA. This is an effective approach, though the poor solids tolerance of such systems increases mud costs due to high dilution rates. Low molecular weight glycols also appear to be effective at controlling dispersion, and have the advantage of not adversely affecting mud rheology.

5 CONCLUSIONS

It is accepted that shale / WBM reactions are extremely complex but continuing research is worthwhile because of the commercial benefits which will come from the use of improved fluids.

Although no mechanisms have yet been proposed which fully explain the behaviour of shales exposed to WBM, we feel that by isolating particular parts of the process — fluid ingress, swelling and dispersion — our understanding of shale problems has advanced. This will result in improvements in mud design, mud engineering and drilling practices.

By understanding the factors which control fluid ingress and shale swelling, we have identified a number of chemical developments which will produce more effective WBM.

The value of this work goes beyond the design and marketing of drilling fluids because of the impact shale instability has on the cost-effective development of oil and gas reserves. The profitability of operators, drilling contactors and a wide range of well-servicing companies is strongly influenced by the quality of wellbores drilled through shales.

REFERENCES

1. G.M. Bol, S.-W. Wong, C.J. Davidson and D.C. Woodland, paper SPE 24975 presented at the SPE European Petroleum Conf., Cannes, France 16–18 November 1992.

2. T.J. Bailey, J.D. Henderson and T.R. Schofield, paper SPE 16525 presented at the SPE Offshore Europe Conf. Aberdeen, 8–11 September 1987.

3. R.C. Minton, in *Chemicals in the Oil Industry: Developments and Applications*, (Ed. P.H. Ogden) p.42. R. Soc. Chem., Cambridge, 1991.

4. D.C. Woodland, *SPE Drilling Engng.,* 1990, **5**, 27.

5. A.H. Hale, F.K. Mody and D.P. Salisbury, paper SPE/IADC 23885, presented at the IADC/SPE Drilling Conf., New Orleans, 18–21 February 1992.

6. M.E. Chenevert, *J. Pet. Tech.* September 1970, 1141.

7. F.K. Mody and A.H. Hale, 1993. paper SPE/IADC 25728, presented at the SPE/IADC Drilling Conf., Amsterdam, 23–25 February 1993.

8. Y.K. Kharaka and F.A.F. Berry, *Geochim. Cosmochim. Acta,* 1973, **37**, 2577.

9. J.P. Simpson, H.L. Dearing and C.K. Salisbury, *SPE Drilling Engng.,* 1989, **4**, 24.

10. M.E. Chenevert and A.K. Sharma, *SPE Drilling and Completion,* 1993, **8**, 28.

11. J.D. Downs, E. van Oort, D.I. Redman, D. Ripley and B. Rothmann, paper SPE 26699 presented at the SPE Offshore Europe Conf., Aberdeen, 7–10 September 1993.

12. T. Ballard, S. Beare and T. Lawless, Paper SPE 24974 presented at the SPE European Petroleum Conf., Cannes, 16–18 November 1992.

13. L. Bailey, J.H. Denis and G.C. Maitland, in *Chemicals in the Oil Industry: Developments and Applications*, (Ed. P.H. Ogden) p.53. R. Soc. Chem., Cambridge, 1991.

14. J.D. Sherwood and L. Bailey, *Proc. R. Soc. Lond.,* 1994, **A 444**, 161.

15. J.M. Cook, G. Goldsmith, T. Geehan, A. Audibert, M.-T. Bieber, and J. Lecourtier, paper SPE/IADC 25729 presented at the SPE/IADC Drilling Conf., Amsterdam, 23–25 February 1993.

16. E. Detournay and A.H.-D. Cheng, *Int. J. Rock Mech. Min. Sci. and Geomech. Abstr.,* 1988, **25**, 171.

17. P.F. Low, in *Clay-water interface and its rheological implications,* (Ed. N. Güven and R.M. Pollastro). CMS workshop lectures, **4**, 157. Clay Minerals Soc., Boulder Co. 1992.

18. K. Norrish, *Faraday Soc. Discuss.,* 1954, **18**, 120.

19. N.T. Skipper, K. Refson and J.D.C. McConnell *J. Chem. Phys,* 1991, **94**, 7434.

20. K. Refson, N.T. Skipper and J.D.C. McConnell, in *Geochemistry of Clay-Pore fluid interactions,* (Ed. D.A.C. Manning P.L. Hall and C.R. Hall) p62. Chapman & Hall, London, 1993.

21. R.M. Pashley, *Chemica Scripta,* 1985, **25,** 22.

22. J.P. Quirk, *Israel J. Chem.,* 1968, **6**, 213.

23. P. Fletcher and G. Sposito, *Clay Minerals,* 1989, **24**, 375.

24. R.P. Steiger, *J. Petr. Tech.* 1982, 1661.

25. S.D. Lubetkin, S.R. Middleton and R.H. Ottewill, *Phil. Trans. R. Soc. Lond.,* 1984, **A311**, 353.

26. P.I. Reid, R.C. Minton and A. Twynam, paper SPE 24979 presented at the SPE European Petroleum Conf., Cannes, 16–18 November 1992.

27. T. Ballard, S. Beare and T. Lawless, in IBC Technical Services Conf. Proc. *Preventing oil discharge from drilling operations — the options*, Dyce, Scotland, 3–24 June 1993.

Pseudo Oil Based Muds – The Outlook

C. A. Sawdon and M. H. Hodder

SCHLUMBERGER DOWELL LIMITED (FORMERLY INTERNATIONAL DRILLING
FLUIDS) C/O ECCI RESEARCH & DEVELOPMENT, PAR MOOR ROAD, PAR,
CORNWALL, UK

1.   Introduction

Pseudo-Oil Based Mud (POBM) may be defined as a drilling fluid, the continuous
liquid phase of which consists of a low toxicity, biodegradable 'oil' which is not
of petroleum origin.

POBMs have evolved as a result of the increasing need to meet difficult drilling
targets with reduced environmental impact. During the 1980s, muds based on low
toxicity mineral oils were heavily used, with the unrestricted dumping of oil mud
coated cuttings into the sea. It was originally thought that little environmental
impact would result by virtue of the very low toxicity of the 'clean oils' to marine
species. However, studies by Davies et al[1] showed that the accumulation of piles
of oily cuttings on the sea bed had a significant impact over at least a 500 metre
radius from the rig. It became clear that the petroleum derived mineral oils did not
biodegrade adequately, especially under the anaerobic conditions within a pile of
cuttings, to allow degradation of the oil and recovery of the sea bed within an
acceptable period.

The need, therefore, arose for biodegradable drilling fluids which still possess the
high performance characteristics of mineral oil based muds. In spite of the strides
made in the development of water based muds for controlling difficult shales,
there are still many occasions when an oil based mud of some kind is necessary.
For example, in high reach deviated wells with long open hole intervals, wellbore
stability (shale inhibition) and lubricity are of paramount importance. As yet,
water based muds do not provide the levels of shale inhibition and lubrication
offered by POBM or mineral OBM. The viability of many offshore field
developments depends on the ability to drill wells, from a single central platform,
with high lateral displacement in order to obtain effective reservoir drainage. For
many such wells, POBM or OBM are viewed as the only effective mud types to
allow the targets to be safely and economically reached.

The increasing legislative restrictions on the discharge of cuttings contaminated with mineral OBM, together with the PARCOM[2] directive on the discharge of 'oils of a petroleum origin', led to the development of synthetic or natural derivative 'Pseudo oils'. These allow effective OBMs to be formulated whilst much enhancing biodegradability and minimizing toxicity.

2.   State of the Art

The following overview is believed to be a fair account, but is based only upon the data and third party reports known to the author at this time. As such, some omissions or minor inaccuracies are probable.

The main pseudo oil types that have been introduced are shown in Table 1, together with some typical properties.

To date, the majority of POBM operations have used ester based fluids, which were first introduced in 1989. Typically favoured are the 2-ethylhexyl esters of C_{12} - C_{18} natural fatty acids (both saturated and unsaturated). These esters display the best available combination of key properties such as kinematic viscosity, flash point, pour point, hydrolytic stability, cost, and biodegradability. Another ester type which was tried caused adverse swelling effects on rubber seals and was withdrawn.

The ester systems suffer from the limitation that at well temperatures in excess of about 150°C the ester will hydrolyse to the component alcohol and fatty acid (especially in the presence of excess lime or cement contamination). This can lead to unstable fluid properties such as large unwanted viscosity increases. No less important is that the alcohol released (typically 2-ethylhexanol) will cause an increase in the marine toxicity of the mud. A second aspect is the relatively high kinematic viscosity (KV) of the esters leading to muds of high plastic viscosity. This leads to poor 'screenability' on shaker screens and difficulty in formulating high density muds because the high concentration of suspended barite only adds to the high plastic viscosity. For the same reason, the use of low oil:water ratios in invert emulsions of brine in ester (high dispersed brine concentration), causes unfavourably high plastic viscosity.

Thus the use of ester muds has been restricted to moderate density and well temperature limits. The high ester contents used lead to more expensive fluids and high levels of ester discharged to the sea with the cuttings.

Table 1:

PSEUDO OIL TYPE			TYPICAL PROPERTIES				
	SG	KV (cSt)	Flash Point (°C)	HTS	Biodegrad. Aerobic	Biodegrad. Anaerobic
Ester	0.85	5 - 6	>150	150°C	Yes	Yes
Polyalphaolefin	0.80	6 - 7	160	High	Yes	No ?
Ether (di-isodecyl ether)	0.83	6	166	High	Yes	No
Acetal	0.84	3.5	>120	180 - 200°C	Yes	Yes
Linear Alkyl Benzene	0.86	4.0	126	High	Yes	?
Rape Seed Oil	0.90	32.5	High	125°C	Yes	Yes
FOR COMPARISON						
Low Tox. Mineral Oil	0.79	3	95+	High	Somewhat	No
Desirable Properties	-	ALAP	AHAP (>120°C)	High (>250°C)	Yes	Yes

KV = Kinematic Viscosity (40°C)
HTS = Hydrolytic Thermal Stability Limit

ALAP = As low as possible
AHAP = As high as possible

Polyalphaolefins (PAOs) typified by oligomers of 1-decene[3], provide high stability drilling fluids of very low toxicity. Again, the kinematic viscosity is relatively high, which dictates the use of high oil:water ratios to minimise plastic viscosity. There is debate on the actual rate of biodegradation of PAOs under the anaerobic conditions in the sea bed cuttings pile. This aspect will be answered more fully in time by properly conducted sea bed surveys.

Di-isodecyl ether has been used and provided drilling muds of high hydrolytic stability. Again, the kinematic viscosity allowed no real viscosity reduction compared to esters. However, the main drawback reported was that anaerobic biodegradation was poor. This is believed to be a consequence of hindrance by the branched alkyl groups and the relative inaccessibility of the ether oxygen to participate in the initiation of the biodegradation process.

More recently, an acetal oil (condensation product of an aldehyde with two moles of an alcohol) has been introduced. Compared to the esters, this displays much reduced kinematic viscosity and increased resistance to hydrolysis under high temperature, alkaline conditions. At present, the acetal is still relatively expensive.

Because of its production in large volumes as a biodegradable surfactant precursor, linear alkyl benzene[4] (LAB) is available at much reduced cost. Compared to esters, the low kinematic viscosity and high stability allow mud formulation to higher densities, high temperatures, and lower oil:water ratios. Contrary to rumour, ther is no tendency for LAB to degrade under downwell conditions to produce free benzene[5] As with PAOs there is uncertainty on the rate of biodegradation of LAB in the sea bed cuttings pile. Twelve field applications have been performed to date and sea bed surveys are due during 1994 to determine the rate of recovery.

For interest, data on rape seed oil is included. Natural triglycerides such as rape seed oil have been mooted on several occasions[6] for use in environmentally benign oil based muds. The attractions are the low cost and ready biodegradability. The drawbacks are (primarily) the high kinematic viscosity and the low well temperature limit of approximately 125°C (because of the ease of hydrolysis). A research project at the Institute of Offshore Engineering (Heriot-Watt University) is studying means of minimising these disadvantages. Low cost muds for less demanding well conditions may be feasible.

3.   Needs for Improvement

Taken as a whole, there is a need to improve POBMs, not only in the performance aspects of hydrolytic stability and plastic viscosity, but more crucially in maximising the rate of biodegradation and recovery of the organically enriched sea bed cuttings pile.

No less important is the need to minimise any human chronic exposure effects such as dermatitis. It is also necessary to avoid swelling effects on nitrile and Viton elastomers which are commonly used in seals or mud motor stators.

Besides the high reach wells mentioned before, high temperature wells requiring stable high density muds are considered an important application for POBM. Both of these well types require the mud's plastic viscosity to be as low as possible. The development of lower viscosity oils would also allow the use of reduced oil:water ratios, with commensurate cost advantages and, importantly, a reduction in the oil concentration discharged with the cuttings (hence more rapid sea bed recovery).

Especially during this era of low and uncertain crude oil prices, the need to reduce drilling costs is strong. Significant reductions in mud unit price should be a target for new POBM systems, as well as providing improved properties to maximise drilling efficiency and minimise lost time.

4.   Environmental Requirements

4.1   Low Toxicity Mineral Oil Based Muds

As the impact of mineral OBM coated cuttings became clearer, increasing restrictions on the allowable level of oil on cuttings (OOC) were introduced on top of the pre-existing toxicity test requirements. For instance, in U.K. waters, an initial (1989) OOC maximum of 150 g/kg was reduced to 100 g/kg, and now (1st January 1994) to 10 g/kg on exploration and appraisal wells. For field development wells, an OOC level of 100 g/kg will be allowed until the end of 1996, when a 10 g/kg limit will be imposed.

In Norway, the current OOC limit for mineral oils is 10 g/kg for all wells. For the Netherlands, no mineral oil discharge is allowed, and in Denmark, mineral oil based muds are not used by general industry consensus. Other countries are expected, sooner or later, to follow the lead set by the European countries.

At present, efforts are being made in Europe to harmonise, for all countries, the toxicity test limits, protocols and species used for both mineral oil muds and POBMs.

4.2   Pseudo Oil Based Muds

The prime addition to the toxicity tests required for mineral oil muds is that the pseudo oil should be biodegradable. The main laboratory test which is recognised is the OECD 301F (28 day) aerobic biodegradation test. Somewhat surprisingly, there is no requirement for an anaerobic biodegradation result, even though anaerobic conditions prevail in sea bed cuttings piles. This is because there is much doubt on the validity of the ECETOC anaerobic biodegradation test. Far more reliance is placed upon the results of sea bed surveys normally carried out twelve months after discharge, even though the grab sampling methods have been criticized. It is anticipated that improved methods will be demanded for both sea bed surveys (e.g. coring of cuttings pile), and for laboratory test protocols.

Sea bed surveys to date on ester mud cuttings piles have shown good clean up and recovery, possibly by a mechanism of quite rapid partial degradation followed by the elution of intermediate, more polar, degradation products. In sea water, these intermediates are believed to biodegrade rapidly and completely under the oxygenated conditions.

There is beginning to be pressure to reduce the level of pseudo oil on cuttings (POOC) discharged. In Norway, the SFT has requested operators to investigate means of reducing POOC. In the UK, the DTI is preparing draft guidelines on the use of POBMs, including an 'ALARA' (as low as reasonably achievable) principle on POOC. In any case, plain economics is an incentive not to discharge expensive pseudo oil.

In Holland, the SSOM have a positive attitude towards POBMs (in contrast to the ban on discharge of mineral OBM). There are no limits for POOC at present. As elsewhere, much importance is attached to the results of sea bed surveys.

Although the requirements in Europe are uncertain and under review, the legislation for offshore USA is far less advanced. For the Gulf of Mexico, it is believed that the EPA has not extended the testing required beyond the 'Sheen Test' and the Mysid Shrimp toxicity test to EPA protocol. The sheen test requires injecting some mud into a bucket of water and viewing whether a surface sheen of oil develops. Candler et al have discussed the needs for clearer EPA guidelines[7].

In overview, much depends upon the results of sea bed surveys, many of which are 'in the pipeline'. One sure thing is that both the drilling industry and the environmental legislation will strive to reduce the impact of POBM cuttings to the sea bed. This will be addressed both by improving biodegradation rates and by reducing the organic loading in the cuttings pile (reducing POOC).

Ultimately, some countries might impose 'zero discharge' (of any kind) regulations which would require the relocation of cuttings to the land (or cuttings re-injection). In this event, there would be little justification for POBMs.

5.   Alternative Technology

The increasing environmental regulations have spurred innovation in the development of alternative muds to mineral OBM, and in the development of cuttings treatment processes to allow the continued use of mineral OBM.

5.1   Alternative Drilling Fluids

Besides POBMs, there has been excellent progress in the development of water based muds (WBMs) with the aim of matching the performance characteristics of oil based mud. The prime benefits obtained from mineral or pseudo oil based muds are as follows:

Borehole stability in shale
High lubricity
Good high temperature stability
Avoidance of stuck pipe (low filtration)
Low formation damage
No corrosion

Many new systems and products have been introduced in recent years to incorporate the above benefits into water based muds. Noteworthy is the introduction of polyol (polyglycerol or polyalkyleneglycol) additives which, when added at circa. 5 to 10% to otherwise conventional WBMs, provide a very valuable increase in shale stability, improve lubricity, and reduce sticking tendencies. Nonetheless, it is inaccurate to claim that OBM performance can truly be matched with current WBMs, and it is difficult to engineer all the attributes into a single WBM formulation.

However, it is thought that ongoing research into improvements in WBM performance, combined with improved drilling hardware (e.g. bits) and practices will allow even the most difficult targets to be reached economically with water based muds (sooner or later).

Another avenue which has received attention[8] is the proposed use of continuous phase organic liquids such as certain glycols which should allow similar drilling efficiency to OBMs but which display significant water solubility. On discharging the cuttings to sea, the mud liquid phase should disperse and dissolve, allowing rapid aerobic biodegradation and much reduced organic loading in the sea bed pile. Reconciliation of this target with desirable low plastic viscosity has proved difficult.

5.2 Cuttings Treatment Processes

(a) Re-Injection

Although not strictly a treatment process, cuttings re-injection is proving a valuable and economical way of disposing of mineral OBM cuttings in an environmentally sound manner. The contaminated cuttings are slurried in water, ground to fine size, and injected into relatively deep formations behind the well casing. Where suitable acceptor formations exist, this technique is most applicable for platform development drilling. Concerns on aquifer contamination or cross-well communication can be pre-empted.

There is currently discussion within Europe (especially Norway and the UK) as to whether cuttings may be transported from one location and injected in another. This question 're-injection or dumping'? will be addressed within PARCOM. If this is allowed, the use of mineral OBM plus cuttings injection will certainly increase, and applications for the more expensive POBMs lessen.

(b) Cuttings Washing

The laundering of OBM cuttings on site with aqueous wash solutions of surfactants has not proved very effective as yet.

(c) Solvent Extraction

A unit to extract oil from cuttings using a volatile, recoverable solvent has been designed and built (essentially 'dry cleaning'). Field trials are imminent.

(d)   Thermal Distillation

A unit is being developed to distil and recover oil from cuttings, using a toroidal shaped fluid bed reactor.

(e)   Other Treatments

Enzymatic degradation of oil has been mooted and improved aqueous washing techniques are under development. Critical fluid extraction using liquid CO_2 has been proposed.

The success of any of the above cuttings treatment processes will depend upon the efficiency, safety, weight, and size of the equipment required. There are many platforms and rigs where this approach cannot easily be applied unless a breakthrough is made in reducing the weight and size of the equipment.

6.   Conclusions

It is clear that the future for pseudo oil based muds depends on many factors. Nonetheless, although many alternative approaches exist or are under development, POBMs will continue to be an important, if specialised, option in the drilling tool kit.

It is believed that improvements in the cost-effectiveness and environmental impact will ensure that POBMs continue to provide an economical and environmentally attractive approach. Ultimately, the combination of new fluids with cuttings processing technology may be required.

This paper has been largely concerned with a North Sea viewpoint. We must not forget the duty we all have to minimise environmental damage globally. Increasing environmental awareness and legislation world-wide will allow POBM's to provide a solution to high drilling performance needs in an ecologically sound manner.

References

(1)   Davies, J.M., et al. Mar. Pol. Bulletin 15 (10): 363-370 (1984)

(2)   PARCOM Decision 88/1 on the use of oil based muds.
       10th Annual REport on the Activities of the Paris Commission. June 1988.

(3)   Friedheim, J. E., et al. SPE 23062 (1991)

(4)   UK Patent Application GB 2258258 A. Brankling, D.

(5)   Personal communication. Ewen, B., Baker Hughes Inteq

(6)   US Patent 4, 374, 737. Larsen, D.E.

(7)   SPE 25993 (1993). Candler et al.

(8)   US Patent 5, 057, 234. Bland, R.G., et al.

Shale Inhibition with Water Based Muds: The Influence of Polymers on Water Transport through Shales

T. J. Ballard, S. P. Beare, and T. A. Lawless

AEA TECHNOLOGY, PETROLEUM SERVICES, WINFRITH, DORCHESTER, DORSET DT2 8DH, UK

1. INTRODUCTION

It is generally accepted that the use of oil-based drilling muds (OBM) provides the most effective stabilisation of reactive shales during drilling. However, environmental legislation reducing the levels of oil on drill cuttings that may be discharged into the North Sea has provided the impetus for research into improving the shale inhibition performance of water-based muds (WBM).

Current methods of shale stabilisation using WBM largely rely on the addition of selected polymeric agents that have been shown to be effective both in field and laboratory trials. However, the operating mechanisms of these additives are not fully understood. The mechanisms by which reactive shales may be stabilised include: chemical stabilisation by reducing dispersive forces, the prevention of water invasion, binding together of shale particles to improve mechanical strength, and increasing mud lubricity (hence reducing mechanical damage from the drillstring).

This paper presents the results of a long-term research project directed towards studying the role of polymers in reducing water invasion into shales, on the premise that if water invasion can be prevented, shale stability will be greatly enhanced. A range of shales has been employed in the study, and initial work was directed towards fully understanding water and ion transport in these rocks to provide a sound basis from which the data obtained in the presence of polymers could be interpreted. This initial data has been published in full elsewhere[2], and hence only a summary is presented here together with a more detailed description of previously unpublished data on saturated salt solution transport processes. The remainder of this paper addresses the influence of individual polymers on water invasion into shales, and then progresses to an examination of selected solids-free mud formulations to identify the presence or absence of synergistic effects between polymers. Mechanisms by which the selected inhibitive polymers operate are proposed, and the product requirements identified as necessary to reduce water invasion into shales are outlined.

2. WATER AND ION TRANSPORT THROUGH SHALES

2.1 EXPERIMENTAL MATERIALS AND METHODS

Shale Samples

Considerable care was taken to select shales appropriate to those encountered in the North Sea. The London Clay and the Oxford Clay are soft gumbo-like mudstones, the Carboniferous shale is hard and compact and selected to represent deep buried shales such as the Kimmeridge Clay. The London clay was cored from a shallow onshore borehole and the Oxford Clay and Carboniferous shale were collected from working quarries. In addition, samples of preserved Tertiary core and Kimmeridge core from the Central Graben have been used. The properties of these rocks are described in Table 1, and their mineralogy in Table 2.

Table 1.   Shale Characteristics

SHALE TYPE	LITHOLOGICAL DESCRIPTION	MOISTURE CONTENT (% of dry weight)	POROSITY (%)	PORE WATER total dissolved solids (ppm)
LONDON CLAY	very soft, no bedding	32.2	46	2,200
OXFORD CLAY	soft, well bedded, fossils	20.1	35	3,100
CARBONIFEROUS	hard, well bedded	6.3	14	3,500
TERTIARY (8185')	hard, bedded, some microfossils	18.7	29	26,500
TERTIARY (5790')	hard, bedded, abundant microfossils	54.6	59	26,500
KIMMERIDGE	hard, well bedded	3.2	8	68,500

The moisture content of each shale was measured, Table 1, and the shales were carefully stored in such a way as to preserve their native moisture content. The composition of the pore water for each shale is also shown in Table 1.

Table 2   XRD Analysis

SHALE	illite/ mica	illite/ smectite	kaolinite	chlorite	smectite	quartz	feldspar	calcite	pyrite
London Clay	38		13	5	15	18	4	2	1
Oxford Clay	25	24	12	5		16	2	4	4
Carboniferous	23	20	13	9		17	8		
Tertiary (8185')	13		20		18	41	2	5	
Tertiary (5790')	10		19		13	43	2	2	2
Kimmeridge	8	11	25			41	4		9

Experimental Procedures

The experimental technique adopted for this study employs radioactive tracers to monitor the transport rates of water and various dissolved ions through the shale specimen. The use of radioactive tracers enables water and ions to be traced without altering the composition of the pore water of the shale. Thus, the role of diffusion and osmosis can be studied without altering the equilibrium state of shale/water systems.

Tracer breakthrough experiments have been used to measure the rate of transport of water and certain ions through shale cores under different experimental conditions (pressure drop, ionic strength and composition). The general principle for all these experiments is the same. A shale core (1.5" diameter, 10-20 mm long, cut such that flow is parallel to bedding) is mounted in a core holder and fluid is circulated passed each end of the core in two separate circulation systems. One reservoir contains the radioactive tracers which are circulated passed the inlet face of the core in an enclosed system. The other outlet or measurement reservoir contains synthetic pore water which is circulated passed the other end of the core in a second separate enclosed system, Figure 1. Monitoring of the outlet reservoir for the appearance and rate of increase in tracer concentration eluted from the core permits calculation of transport rates[1,2]. All eluent concentration data is normalised to the inlet concentration, thus permitting easy comparison between experiments and tracers.

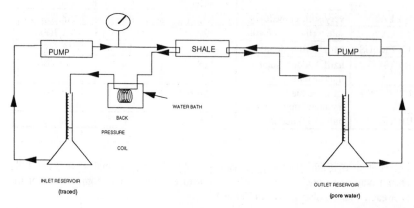

Figure 1 Experimental Set-up for Radiotraced Experiments
(the back pressure coil is removed for diffusion experiments)

2.2   RESULTS AND DISCUSSION

2.2.1   DIFFUSION EXPERIMENTS

In these experiments there was no pressure drop across the shale core, therefore any transport process would be diffusion. Most experiments have been performed tracing tritium (for water) and chloride ions, although, the diffusion rates of sodium and calcium have also been measured in the London Clay. Experiments have also been performed in which the composition of the water contacting the shale has been changed, in order to observe any osmotic phenomena.

Diffusion of Pore Water

Diffusion experiments tracing water and chloride ions in pore water have been performed on all the shales used in this study and Figure 2 is an example of the tracer profiles obtained in these experiments. For all the shales, the diffusion rate for water is higher than that for chloride ions, Table 3. The reason for the difference in rates between chloride ions and water is not immediately clear. The free diffusion rate of chloride ions is lower than that for water, but this difference is less than that observed through the shale samples. It may be that charge effects on the pore walls of the shale further reduce the diffusion rate. There is a correlation between shale porosity and diffusion rate of water (and to some extent chloride ions), with diffusion rate increasing with increasing porosity, Figure 4. At very low porosities surface interactions between ions and shale minerals dominate. As porosity increases surface interactions become less important compared to free diffusion within the pore spaces, and rates of diffusion increase as porosity increases until the free diffusion value is reached (at 100% porosity).

Table 3   Summary of Diffusion Rates

SHALE	FLUID	Diffusion rate WATER $(10^{-10})m^2s^{-1}$	Diffusion rate CHLORIDE $(10^{-10})m^2s^{-1}$	Diffusion rate POTASSIUM $(10^{-10})m^2s^{-1}$	Diffusion rate SODIUM $(10^{-10})m^2s^{-1}$	Diffusion rate CALCIUM $(10^{-10})m^2s^{-1}$
LONDON CLAY	Pore water	1.9	0.59		3.41	4.62
	10% pore water	1.8	0.23			
	10% pore water	1.81	0.24			
	2% KCl	1.7	1.2	1.1		
	5% KCl	1.7	1.5	1.2		
	10% KCl	1.7	1.6	1.3		
OXFORD CLAY	Pore water	1.9	1.2			
	10% pore water	2.0	0.56			
CARBON-IFEROUS	Pore water	0.23	0.09			
TERTIARY 8185'	Pore water	1.25	0.24			
	1% pore water	1.24	0.12			
TERTIARY 5790'	Pore water	3.05	1.9			
	1% pore water	2.88	1.2			
Kimmeridge	Pore water	0.17	<0.05			
	10% pore water	0.17	<0.05			

The diffusion of sodium and calcium ions through the London Clay has also been studied (Figure 3). For both cations there is a delayed breakthrough into the outlet reservoir, this is due to ion exchange of the tracers with sites on the clay surfaces. It can be inferred from the longer breakthrough time for calcium ions that there is more exchange potential for calcium within the London Clay than sodium. Unlike chloride ions, once a steady state flux across the core has been achieved, the diffusion rate for both cations is greater than water. This

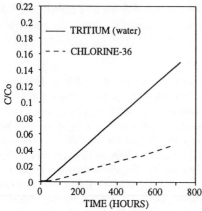

Figure 2. Diffusion rates of water and chloride ions
in the London Clay

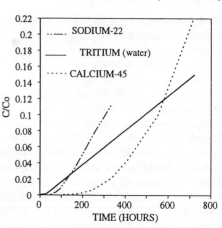

Figure 3. Diffusion rates of sodium, calcium and
and water in the London Clay

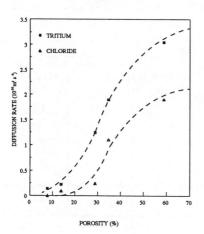

Figure 4. The Effect of Porosity on Diffusion Rate
in Bedded Shales

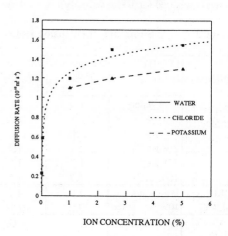

Figure 5. The Effect of Ion Concentration on Diffusion Rates
in the Oxford Clay

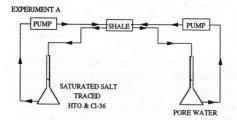

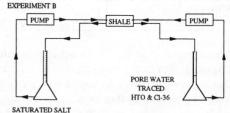

Figure 6. Experimental set-up for two Experiments, Designed to Trace
the Diffusion of Chloride ions in Both Directions from a Saturated Salt Solution

phenomenon has been attributed by several workers to the existence in sorbed ions of a surface diffusive flux[3,4], whereby ions diffuse along mineral surfaces in addition to diffusion within the pore water.

Diffusion from Diluted Pore Water

The existence of osmosis in water/shale systems has been investigated. A solution less saline than pore water (diluted pore water) was exposed to one side of the shale core, whilst the other end (outlet end) of the core was contacted with pore water. In these experiments diffusion rates for water and chloride ions were initially obtained from pore water in order to provide baseline data prior to contacting the shale with the diluted pore water. In this way a direct comparison of water transport rates could be obtained on the same core. If osmosis were to occur then the rate of water transport through the shale core from the dilute reservoir to the pore water reservoir would be faster than the initial pore water diffusion rate.

The results show (Table 3) that, in all the shales tested, there is no discernible difference in water transport rates between diffusion from pore water and diluted pore water. These results indicate the absence of osmosis as an important mechanism of water transport in these shale/water systems. However, in the case of the chloride ion, diluting the pore water always reduced the rate at which the ion diffused through a particular shale.

Diffusion from Potassium Chloride Solutions

Experiments have been performed on the London Clay where one side of the core was contacted with various concentrations of potassium chloride solutions (2%, 5% and 10%) traced with tritium and chlorine-36. The potassium build-up in the outlet reservoir was monitored by ICPOES (Inductively Coupled Plasma Optical Emission Spectroscopy).

The diffusion rates for water are similar to those observed in the pore water experiments and seem to be independent of KCl concentration, Table 3. The chloride diffusion rates, on the other hand, increase with increasing chloride concentration to a maximum (when the chloride concentration is 2.5%) approaching the diffusion rate of water. Figure 5 shows the influence of chloride ion concentration on the diffusion rate (this graph includes data from the diluted pore water experiments) and indicates that the diffusion of chloride ions is strongly influenced by concentration gradients. Potassium is initially delayed, but then diffuses at a similar rate to chloride ions.

Diffusion from Saturated Salt Solutions

In the previous experiments concentration gradients were found to be an important factor in controlling ion diffusion rates. However, the changes in test water composition did not introduce any significant water concentration gradient across the shale. Two experiments using saturated salt (NaCl) solutions were performed to investigate water diffusion where there is a significant water concentration gradient between the test fluid and the shale pore water. This concentration gradient may have a significant effect on the transport of water.

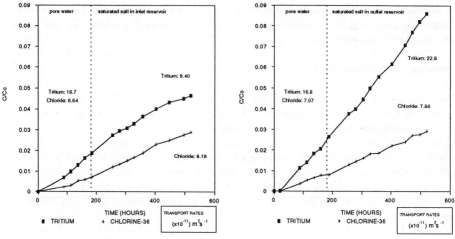

Figure 7. Saturated salt diffusion (experiment A). Tracers placed in the saturated salt reservoir.

Figure 8. Saturated salt diffusion (experiment B). Tracers placed in the outlet (pore water) reservoir.

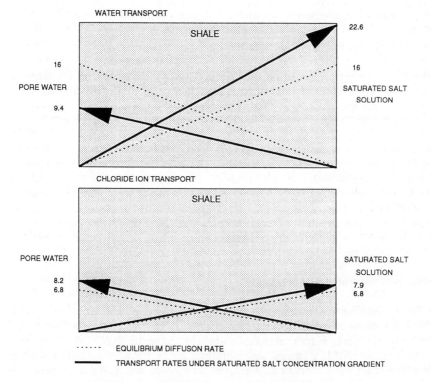

Figure 9. Schematic diagram summarising water and chloride ion transport rates through the Oxford Clay which is contacted with pore water on one side and saturated salt solution on the other.

The saturated salt diffusion experiments were performed using the Oxford Clay. Pore water diffusion rates were measured first, before placing a saturated salt solution at one end of the core with pore water at the other. Two experiments were carried out so that the transport of water and chloride ions could be measured in both directions through the shale. In one experiment (experiment A) tritium and chlorine-36 tracers were placed in the saturated salt solution to measure the rate of transfer of both ions from the saturated salt to the pore water. In the other experiment (experiment B) the tracers were placed in the pore water reservoir to measure rates of transport to the salt solution, Figure 6. The volumes of the reservoirs were also monitored to check for any bulk transfer of fluid.

In both experiments there was a small but measurable (0.2 cm^3/day) bulk transfer of fluid from the pore water reservoir to the saturated salt reservoir. This volume transfer is equivalent to a pressure drop of 6 psi across the core. The rate of water transport through the shales is in agreement with the volume changes observed, Figures 7 and 8. In experiment A (tracers in the saturated salt reservoir) the rate of water diffusion decreases when the inlet reservoir is changed to saturated salt, because the mass transfer of water is in the opposite direction to the measured tracer rate. In experiment B where the outlet reservoir contains saturated salt (tracers in the pore water reservoir) the diffusion rate for water increases due to the volume transfer of fluid into the salt reservoir.

In the case of chloride ions, where the transport is measured in the same direction as the concentration gradient, i.e. where the saturated salt reservoir is traced, (experiment A), the rate of chloride diffusion increases to the rate of water diffusion in this direction (which has, of course, decreased). So that after the introduction of the salt reservoir, the chloride ions and water travel at more or less the same rate which is higher than the equilibrium (pore water) chloride diffusion rate, but, lower than the equilibrium water diffusion rate.

Where the diffusion rate of chloride ions is measured against the concentration gradient, i.e. where the pore water reservoir is traced (experiment B), one would expect a marked decrease in the chloride ion diffusion rate as observed in the previous osmosis-type experiments. However, in this experiment the chloride rate remained largely unaffected and if anything, a marginal increase in rate was observed. A possible explanation is that the chloride ions are carried along with the mass transfer of water, despite the negative concentration gradient.

Figure 9 is a schematic diagram summarising the results of water transport from these experiments. The rate of water transport to and from the saturated salt reservoir is shown with the normal pore water (equilibrium) diffusion rates. The transport of water to the saturated salt reservoir is about three times the diffusion rate in the other direction. This imbalance in transport rates causes the observed mass transfer. The normal equilibrium diffusion rate lies midway between these rates. So that the amount of rate increase (over pore water diffusion) in the saturated salt direction is equivalent to the rate decrease (compared to pore water diffusion) in the opposite direction. In both systems (pore water-shale-pore water and pore water-shale-saturated salt) the total diffusion rate for water in both directions is the same (i.e. 16+16=32 and 22.6+9.4 =32).

Figure 9 also summarises the chloride transport results. It can be seen that there is no significant net transfer of chloride from one reservoir to the other as the transport rates are similar in both directions. Both rates are slightly higher than the pore water diffusion rate. The increase in rate towards the pore water is due to the chloride concentration gradient which increases the diffusion rate to that of the water diffusion rate.

The cause of the transfer of chloride ions in the other direction i.e. against the gradient is less obvious. However, in this direction, the rate of water transport is greater than the equilibrium diffusion rate, the increase is caused by the observed mass transfer. In previous experiments we have found that during mass transfer (i.e. advection) chloride ions and water travel at the same rate. Therefore, although chloride ions would not be expected to diffuse against the concentration gradient, the mass transfer of fluid will carry the chloride ions at the same rate as the water. Thus, the rate of transport of chloride ions is the same as the rate of mass transfer of water. The rate due to the mass transfer of water is the total rate minus the equilibrium diffusion rate (i.e. 22.6 - 16) which gives a rate of $6.6 \times 10^{-11} m^2 s^{-1}$ which is in agreement with the observed rate (average $6.8 \times 10^{-11} m^2 s^{-1}$).

2.2.2 INFLUENCE OF APPLIED PRESSURE (ADVECTION) ON WATER TRANSPORT RATES

A pressure drop was applied to the inlet face of the shale core by means of a narrow bore tube to create the desired back pressure. In this way the influence of applied pressure (advection) on water transport rates could be investigated. The rate of water transport increases linearly with increasing pressure. The rate of transport of chloride ions also increases linearly with pressure and gradually approaches the water rate as advective flow takes over from diffusion. A pressure of 145 psi is sufficient to induce fully advective flow in the Oxford Clay, Figure 10. Table 4 summarises advection rates at 300 psi applied pressure for all the shales studied. There is no simple relationship between the diffusion rate of water and advection rates in a shale. The magnitude of increase in rate from zero pressure (diffusion) to 300 psi applied pressure varies depending on the shale type. As with diffusion rates there is a wide variation in rates for different shales.

Table 4. Summary of Water Advection rates and Shale Permeabilities

SHALE	Advection Rate (300psi) $(10^{-10})m^2 s^{-1}$	Permeability (nd)
OXFORD CLAY	10.00	500
CARBONIFEROUS	1.90	70
TERTIARY (8185")	3.40	99
TERTIARY (5790")	95.00	6,000
KIMMERIDGE	0.24	12

2.2.3 WATER AND ION TRANSPORT THROUGH SHALES: SUMMARY AND CONCLUSIONS

Transport From Pore Water

Rates of water transport through shales vary considerably depending on the nature of the shale, Table 3. The diffusion rates for the high porosity shales can be greater than water advection rates (at 300 psi) for more compact shales (Oxford Clay c.f. Carboniferous shale/Kimmeridge Clay). There is a correlation between porosity of the shale and the diffusion rate of water through it, Figure 4, and this relationship also applies to some extent to the diffusion of chloride ions in pore water. The diffusion rates of chloride ions through shales from pore water are always lower than that for water, the rates appear to depend on rock fabric as well as porosity. Exchangeable cations such as sodium and calcium, after an initial lag period, diffuse faster than water from pore water solutions.

Applying a pressure always increases the water transport rates through shales, though the magnitude of rate increase above that of diffusion depends on the particular shale; there is no simple relationship between diffusion rates and advection rates.

Effect of Concentration Gradients on Diffusion through Shales

The controlling fluid parameter for determining diffusion rates through a given shale is concentration gradient. When there is a concentration gradient to chloride ions, the rate of chloride ion diffusion, in this direction, increases to that of the water rate. There is a reduction in the rate of chloride ion diffusion against the concentration gradient, resulting in the net transfer of chloride ions to the less saline solution. After a certain concentration there is no further increase in the diffusion of chloride ions as chloride ion diffusion is restricted to the diffusion rate of water.

When there is a significant concentration gradient across the shale to water (i.e. when saturated salt contacts the shale) the transport of ions such as chloride is complicated by the preferential diffusion of water in the opposite direction to the ionic concentration gradient. Tracing the movement of water and chloride ions in both directions gives a full picture of ion and water movement both into and out of the shale from the saturated salt solution. It was found that water travels at a higher rate (than pore water equilibrium diffusion) from the pore water to the saturated salt reservoir, causing a small amount of volume transfer into the saturated salt solution. There is diffusion of water in the other direction but this is lower than the equilibrium diffusion rate. Chloride ions are transported in both directions, but, as expected, the rate of diffusion from the saturated salt to the pore water is higher than the measured equilibrium diffusion rate. This rate is the same as the water in this direction, which suggests that the rate of chloride ion diffusion is restricted to the rate of water diffusion and that unlike sodium and calcium, chloride ions cannot travel through shales faster than water.

Chloride ion diffusion without a concentration gradient is slower than the water diffusion rate through shales. Imposing a concentration gradient w.r.t. chloride ions increases the diffusion rate (to that of the water rate) in the direction of the concentration gradient (e.g. 5% KCl), causing a net transfer of chloride ions to the less saline solution. At very high salt concentrations, the water concentration will be significantly reduced which creates a water concentration gradient in the opposite direction. The effect of this, is to reduce the rate of water diffusion in the direction of the chloride concentration gradient. This means that the chloride diffusion rate is also reduced because chloride ions cannot diffuse faster than water. In the saturated salt experiment performed on the Oxford Clay the chloride ion transport caused by mass transfer of water into the saturated salt reservoir was the same as the chloride diffusion rate in the other direction resulting in no net transfer of chloride ions despite the large concentration gradient. In this situation, a net transfer of chloride can only occur when the concentration gradient of water is sufficiently reduced (by transfer of water into the saturated salt solution) so that the diffusion rate out of the saturated salt solution is higher than the mass transfer into this solution.

It is clear from these experiments that shales do not act as semipermeable membranes as ions are free to diffuse into and out of shales, the rates for a particular shale being controlled by concentration gradients. Mass transfer of water out of a shale only occurs where the solution contacting the shale has a very high salinity causing a concentration gradient to water, however, this phenomenon is readily explained in terms of diffusion.

3.   INFLUENCE OF POLYMERS ON WATER INVASION INTO SHALES

Having obtained baseline data on water transport through shales, the influence of polymers on water invasion was investigated. Four commercially available "inhibitive" polymers were used, PHPA (partially hydrolysed polyacrylamide), PAC (polyanionic cellulose), a polyamine and a glycol derivative, together with xanthan (a viscosifier) and a fluid loss agent (starch). Those polymers which caused rate reductions on the Oxford Clay were tested on the Tertiary cores.

All of the polymer advection experiments followed the same basic procedure. Initially, the rate of flow of the base fluid (sea water plus 5%KCl) through the shale was measured. Then the transport rate of fluid from the polymer solution was measured under the same conditions on the same shale core so that a direct comparison of the transport rates could be made. An applied pressure of 300 psi was used for all experiments. In most experiments tritium and chlorine-36 were used to monitor transport rates, in addition, to measuring the volume transported through the shale core. In some instances where the shale permeability was too high or the viscosity of the test fluid was very high, tracers were not used and only the volume flow through the core was measured.

3.2   RESULTS AND DISCUSSION

The results from experiments investigating the effect of single polymers on water transport rates through shales are summarised in Table 5.

Xanthan

Oxford Clay - Both 0.5 ppb and 1.5 ppb xanthan were not found to affect the transport rate of water through the Oxford Clay, Table 5. There is some evidence that some fragments of the xanthan molecule can penetrate the shale core, however, xanthan is not strongly adsorbed onto shales.

PAC

Oxford Clay - The transport rate of water through the Oxford Clay was not significantly affected by either 0.5 ppb (pounds per barrel) or 2 ppb PAC, Table 5. Analysis from the experiment showed that the polymer had not invaded the shale beyond the first millimetre. An additional experiment using carbon-14 labelled CMC provided supporting evidence that these polymers cannot invade the Oxford Clay.

Table 5 Summary of Single Polymer Advection Resutls

POLYMER	CONCENT-RATION (ppb)	SHALE	RATE BEFORE POLYMER (cm^{3}hr^{-1})	RATE AFTER POLYMER (cm^{3}hr^{-1})	RATE REDUCTION (%)
XANTHAN	0.5	OXFORD	0.254	0.262	
XANTHAN	1.5	OXFORD	0.215	0.222	
PAC	0.5	OXFORD	0.244	0222	9
PAC	2.0	OXFORD	0.237	0.222	7
PHPA	2.0	OXFORD	0.287	0.214	25
PHPA (mw700.000)	2.0	OXFORD	3.01	0.180	40
PHPA	0.5	TERTIARY 8185'	0.0566	0.048	15
PHPA	2.0	TERTIARY 5790'	2.51	1.01	60
STARCH	2.0	OXFORD	0.304	0.209	31
STARCH	8.0	OXFORD	0.265	0.207	22
STARCH	2.0	TERTIARY 8185'	0.158	0.054	66
STARCH	3.5	TERTIARY 5790'	2.39	0.717	70
POLYAMINE	2.0	OXFORD	0.251	0.131	48
POLYAMINE	8.0	OXFORD	0.310	0.161	48
POLYAMINE	3.5	TERTIARY 8185'	0.0955	0.031-0.009	66-90
GLYCOL	3.0	OXFORD	0.284	0.203	26
GLYCOL	10.0	OXFORD	0.304	0.163	46
GLYCOL	10.0	TERTIARY 8185'	0.0496	0.033-0.016	34-67
GLYCOL	10.0	TERTIARY 5790'	2.68	1.8	32

PHPA

Oxford Clay - The effect of the molecular weight of PHPA on the reduction of water transport rate has been tested on the Oxford Clay. One experiment used PHPA (2 ppb), molecular weight 6-7 million, whilst 2 ppb of a PHPA with a molecular weight of 700,000 was used in another experiment. It should be noted that the molecular weights quoted are the initial molecular weights, PHPA is very sensitive to shear degradation and the process of making up the polymer solution will degrade it. The results (Table 5) reveal that PHPA reduces water rates through the Oxford Clay, the higher molecular weight polymer by 25% and the lower molecular weight PHPA by 40%. PHPA was detected in the effluent from these experiments (presumably material of low molecular weight) indicating that a proportion of the polymer can invade the Oxford Clay. PHPA is quite strongly adsorbed onto shales.

Tertiary - The effect of PHPA on water transport in both samples of preserved Tertiary core has also been investigated. A slight reduction in water rate of 15% was observed in the low porosity Tertiary core, Table 5. With the high porosity Tertiary core, on the other hand, PHPA caused a dramatic reduction in invasion rate of 60%.

These results on the effect of PHPA with different shales give an insight into the mechanism causing the reduction in water invasion rates. It appears that PHPA is more effective where the polymer is more likely to be able to invade the rock. For instance, a greater rate reduction is observed in the Oxford Clay with the lower molecular weight PHPA than with the conventional PHPA. PHPA is least effective with the low porosity Tertiary core where invasion is most unlikely, and most effective with the high porosity Tertiary core where a greater proportion of the polymer would be expected to invade the shale. Therefore, reduction in water invasion depends on the ability of the polymer to invade the shale, which suggests the mechanism is adsorption and retention of the PHPA within the shale which will cause pore restrictions and possibly blocking.

Starch

Oxford Clay - 8 ppb starch caused a decrease in the transport rate of water of about 20% in the Oxford Clay. A slightly greater decrease in rate was observed with 2 ppb starch (Table 5), however for this experiment the initial transport rate through the shale was higher than usual which may indicate the presence of a micro fracture. In both tests starch was not detected beyond the first millimetre of the inlet face.

Tertiary - Starch caused a considerable reduction in water rates (65-70%) in both preserved Tertiary core samples. Unexpectedly the magnitude of the rate reduction was similar for both shale samples, Table 5. However, the initial rate for the low porosity Tertiary was unusually high suggesting either a microfracture or a high permeability zone. 2 ppb starch reduced this rate to one more characteristic for this Tertiary core (compare with the PHPA test, Table 7).

The effect of starch on the different shales is rather less predictable than the other polymers tested. In some instances high rate reductions can be achieved where the initial permeability, of the shale concerned, was unusually high. The is no evidence that even a proportion of the polymer can travel through the shales (unlike PHPA). It appears that the starch plugs large pores and fractures on the face of the shale with only minimal invasion into the pore networks of the rock.

Polyamine

Oxford Clay - The water transport rates through the Oxford Clay is halved in the presence of 8 ppb polyamine. This is similar to the result obtained with 2 ppb polyamine, Table 5.

Tertiary - The polyamine caused the largest decrease in water rate through the low porosity Tertiary core, Figure 11. The transport rates of water and chloride ions were immediately reduced to approximately 30% of the initial rate and continued to decrease to 10% of the original rate 300 hours after the polymer addition.

The polyamine, being cationic, adsorbs very strongly to shales, it has a relatively low molecular weight (approximately 50,000) which allows invasion into the Oxford Clay and probably the low porosity Tertiary core (though depth of invasion has not been measured on the Tertiary). The polyamine is adsorbed so strongly by the Oxford Clay that even after 500 hours there is no sign of the polymer in the effluent (i.e. all of the polymer forced into the shale has adsorbed). The depth of invasion was detected by sectioning the core and measuring the zeta potential of each core section. Invasion adsorption and blockage (which may occur through an alteration of the fabric of the shale) would appear to be the mechanism by which this additive slows water invasion into shales. The narrower pore throats of the low porosity Tertiary shale allows the polyamine to reduce the water invasion rates initially by 60% then eventually (after 300 hours) the invasion of water almost stops.

Glycol Derivative

Oxford Clay - 3% glycol eventually reduced the water rate by 25% though there was some time lag between the addition of the glycol and the decrease in the water rate. A 10% glycol solution caused a 50% rate reduction. The viscosity of the effluent collected directly from the cores indicate that the glycol was transported through each shale core.

Tertiary - A 10% glycol initially has little effect on the transport rate of water through the low porosity 8185' Tertiary core, Figure 12, but with time the rate gradually decreases to 33% of the original rate after 600 hours which gives a rate reduction of 46% averaged over the whole test. The viscosity of the effluent collected after this test was only slightly greater than the base fluid viscosity indicating that the glycol cannot pass unrestricted through this rock (unlike with the Oxford Clay). A 32% rate reduction was observed with a 10% glycol solution on the high porosity Tertiary core. In this instance the effect was immediate, and the effluent viscosity revealed that the whole glycol solution can pass through this shale.

The magnitude of the rate reduction in the high porosity (5790') Tertiary shale is directly attributable to the increase in viscosity of the effluent (from Darcys' equation). Therefore, in this instance the reduction in water rate is purely a function of viscosity and there has been no alteration in the permeability of the shale. The magnitude of the rate reduction caused by the glycol derivative is related to the concentration, unlike the other polymers tested. This supports this proposed mechanism of viscosification of the filtrate within the shale. However, in the case of the Oxford Clay the observed rate reduction is higher than that expected from Darcy flow. However, Darcy flow assumes no rock/fluid interactions. In such shales as the Oxford Clay the effect of viscosity on flow rate at a given pressure drop may be greater due to the narrow pores.

3.3   INFLUENCE OF POLYMERS ON WATER INVASION INTO SHALES: SUMMARY AND CONCLUSIONS

In order to reduce water invasion some interaction between the polymer and the shale is necessary which requires some degree of invasion of the polymer. The high permeability and porosity of the Tertiary core from 5709' will facilitate greater degrees of invasion from high molecular weight polymers and both PHPA and starch give greater rate reductions in this shale than with the Oxford Clay If the polymer size is too large compared to the shale pore sizes e.g. PAC and Oxford Clay, the polymer cannot invade and there is no change in the rate of water invasion. At the other extreme, if the polymer is small enough to pass straight through the shale unimpeded, then the observed rate reduction of water invasion is related to the viscosity of the filtrate, e.g. glycol and high porosity Tertiary core. Between

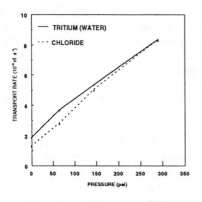

Figure 10. The Effect of a Hydraulic Gradient on Water
Chloride Transport Rates through the Oxford Clay

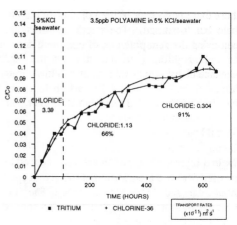

Figure 11. The Effect of Polyamine on Water Transport
through the Tertiary (8185') Shale.

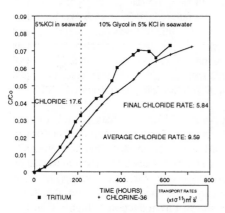

Figure 12. Effect of Glycol on water and chloride transport
through the Tertiary (8185') shale, at 300psi

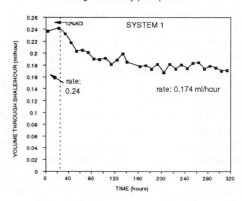

Figure 13. Effect of system 1 on water transport
through the Oxford Clay

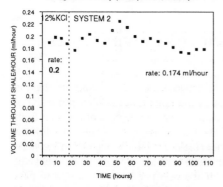

Figure 14. Effect of system 2 on water transport
through the Oxford Clay

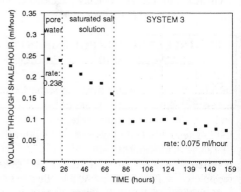

Figure 15. Effect of system 3 on water transport
through the Oxford Clay

these extremes where the polymer is small enough to enter the first few millimetres of the shale but too large to pass right through the core, this retention leads to greater rate reductions than simply viscosifying the filtrate (compare the glycol on both Tertiary cores). It may be possible to predict the effect of a particular polymer on water invasion rates if the pore size distribution of the shale and the hydrodynamic radius of the polymer is known.

4   EFFECT OF POLYMER MIXTURES ON WATER TRANSPORT THROUGH SHALES

Having obtained baseline data on single polymer systems (under an imposed 300psi pressure gradient) it should be possible to assess the presence of any synergistic effects with multi-component systems similar in composition to water based muds (but without any solids). Three systems all containing a mixture of polymers have been tested on the Oxford Clay. The complete formulations of the systems are shown in Table 6.

Table 6   Formulations of Multi-component Systems

SYSTEM 1	KCl	12%
	PHPA	1.4ppb
	PAC (low viscosity)	2.8ppb
	xanthan	0.5ppb
SYSTEM 2	KCl	12%
	glycol derivative	3%
	fluid loss agent	5.25ppb
	xanthan	1.4ppb
SYSTEM 3	NaCl	126ppb
	fluid loss agent	5.6ppb
	xanthan	1.05ppb

System 1 - KCl/PHPA/PAC Formulation

The profile showing flow rate through the core is shown in Figure 13. The initial flow rate through the core with the base solution (12% KCl) was 0.24 cm^3/hr. This was gradually reduced to 0.174 cm^3/hr after changing to system 1 and represents a rate reduction of 25%, (Table 7). At the end of the experiment the effluent was collected and its viscosity measured; there was no increase in viscosity over the base solution. A noticeable slimy residue was observed on the face of the core upon removal from the core holder.

TABLE 7.   Summary of Water Transport Rates from Multicomponent Systems (OXFORD CLAY)

SYSTEM	BASE BRINE RATE (cm^{3}hr^{-1})	SYSTEM RATE (cm^{3}hr^{-1})	RATE REDUCTION (%)
SYSTEM 1	0.24	0.178	25
SYSTEM 2	0.20	0.174	13
SYSTEM 3	0.24	0.075	70

Remarkably, the rate reduction from this experiment was the same as that obtained by PHPA on its own with the Oxford Clay (Table 3). Thus, no additional benefit from the presence of xanthan or PAC (which on their own do not affect water invasion rates) was apparent. The presence of the slimy polymer residue/filtercake on the face of the core may improve shale stability in the field, for instance, by increasing the lubricity of the shale surface and preventing dispersion of the outside layer of clay particles.

System 2 - KCl/Glycol/Formulation

The profile showing flow rate through the Oxford Clay in the presence of system 2 is shown in Figure 14. The flow rates through the core are rather variable and only a marginal rate reduction of 13% was observed 80 hours after contact with system 2. The effluent from this core had an increased viscosity over the base solution of 8% indicating that a proportion of the glycol passed through the core.

When compared to the result of the experiment using just 3% glycol (on its own) with the Oxford Clay (which gave a rate reduction of 25%), the effectiveness of the glycol in the formulation of system 2 seems to be diminished. However, there was a significant time lag before the full effect of the glycol is realised. Over the same time span the rate reduction for glycol on its own and system 2 is the same. This indicates that the rate reduction caused by system 2 can be directed attributed to the glycol. As with system 1 there are no additional synergistic effects from the other components in the formulation. The viscosity of the effluent reveals that only a proportion of the glycol had passed through the core in this instance, although from previous work it was found that virtually all the glycol could pass through the Oxford Clay. Again this is probably a result of the time length of the experiment because the rate reduction is dependent on the viscosity of the filtrate which requires time to saturate the core.

System 3 - Saturated Salt Formulation

In this experiment the flow rate of pore water through the shale was measured initially, which gave a rate of 0.238 cm^3/hr. The effect of a saturated salt solution (no polymers) was measured before introducing system 3. It would appear from the data (Figure 15) that though the rate sharply decreased (to 0.185 cm^3/hr) it had not stabilised before changing to system 3. The switch to system 3 caused an abrupt decrease in rate which eventually stabilised to 0.075 cm^3/hr. The effluent from the core had a viscosity of almost 2 cP which was double the viscosity of the pore water. At the end of the experiment there was a noticeable build-up of polymer on the shale surface.

This formulation was the most effective at slowing water invasion into the Oxford Clay causing a reduction of 70%. The main mechanism appears to be the increase in filtrate viscosity produced by the dissolved salt. According to Darcys' Law a doubling of viscosity should reduce the flow rate through the core by half, which does not account for the observed rate reduction. A small portion of the rate reduction will be due to preferential diffusion of water into the saturated salt solution (section 2.2.1). A further contribution to the rate reduction may be due to the fluid loss agent although its presence in system 2 did not seem to enhance the performance of the glycol.

5.   CONCLUSIONS

In order to reduce water invasion into shales, the additive/polymer must be able to invade the porous matrix or microfractures of the shale. There are two potential mechanisms which can cause retardation of water invasion:

(i).   At least partial invasion of the polymer into the shale, followed by adsorption and/or blocking of pore throats/fractures to reduce the permeability of the shale.

(ii).   Viscosification of the filtrate in the shale caused by the polymer/additive travelling through the shale within the water phase.

The size of the additive/polymer in relation to the pore size (or fracture size) of the shale is the most important factor in predicting the effect on water invasion into the shales.

In the two multi-component systems comprising a shale stabilising polymer with a fluid loss agent and a viscosifier, there was no evidence of synergy between the additives. The reduction in invasion rates was the same as that observed for the shale stabilising polymer alone. The largest reduction in invasion rate, produced by the saturated salt/fluid loss agent/viscosifier combination, was mostly attributable to the filtrate viscosity increase caused by the dissolved salts.

6.   SOME CONSIDERATIONS FOR WELLBORE SHALE INHIBITION

This study has shown that although water invasion rates can be reduced, timescales are usually long (a few days) and the invasion is only partially retarded. Furthermore, additives which have been shown to work in the field do not necessarily influence water invasion rates (e.g. PAC). It is therefore clear that controlling water invasion, although important, is not the only mechanism by which inhibitive polymers operate.

A second mechanism may result from polymer adsorption on the shale surface. Polymers typically form a slippery film which would aid lubricity, and reduce mechanical damage. Strongly adsorbed polymers may also bind the shale surface together, reducing dispersion.

ACKNOWLEDGEMENTS

This study formed part of a long term research project sponsored by: Norsk Hydro, AGIP (UK), Amoco, Bow Valley (UK), BP, Fina, Kerr-McGee, LASMO, Mobil, and Statoil. Amoco and Mobil supplied the preserved core. Initial discussions with Nick Jefferies (AEA Technology, Harwell) on radionuclide transport through mudrocks were very helpful.

REFERENCES

1.   Gilling, D., Jefferies, N. L. and Lineham, T. R. - An Experimental Study of Solute Transport in Mudstones, Harwell Report , AERE - R12809, NSS - R109, 1987.

2.   Ballard, T. J., Beare, S. P. and Lawless, T. A. - Fundamentals of Shale Stability: Water Transport through Shales, paper SPE 24974, presented at 1992 European Petroleum Conference, Cannes, France, November 16-18, 1992.

3.   Rasmusson, A. and Neretnieks, I. - Surface Migration in Sorption Processes, SKB Technical Report 83-37, 1983.

4.   Skagius, K. and Neretnieks, I. - Diffusion Measurements of Cesium and Strontium in Biotite Gneiss, SKB Technical Report 85-15, 1985.

Development and Application of Cationic Polymer Drilling Fluids for Shale Stabilization

J. Dormán and É. Banka

HUNGARIAN OIL AND GAS PLC, OIL AND GAS LABORATORIES, BUDAPEST, HUNGARY

INTRODUCTION

Maintaining borehole stability while drilling is a major problem when the bit encounters water sensitive shale formations. This action helps to optimize other oilwell operations and reduce overall drilling cost. Composition and chemistry of the applied drilling fluid play primary role in borehole stabilization.

Highly adsorptive, high molecular weight synthetic polymers are known to be the best candidates for solving shale related problems. Their efficiency, which is based on "encapsulating" (or coating) effect can be enhanced further by the addition of specifically adsorbed inorganic cations (i.e. potassium, ammonium, etc.).

Most of the polymers used for this purpose were selected from the anionic and, or non-ionic types. Recent developments in drilling fluid chemistry and convertible experiences from other industries (water treatment, sludge filtration/phase-separation) led to formulation of new cationic polymer drilling fluids.

However the extremely efficient cationic polymers are incompatible with most of the additive used in the conventional drilling fluid systems. This effect makes their use very difficult in new drilling mud formulations.

Predominate requirement when penetrating pay zone of oil and gas wells is the prevention of formation damage, or preservation of original formation parameters. This becomes even more obvious, and more difficult on the other hand in case of horizontal wells. Most of the sandstone reservoirs (in Hungary) involve interlayered shale sections and intergranular shale(clay) contaminants.

Long term stabilization of these sections strongly affects the extent of formation damage. Properly designed drilling mud chemistry has come to the highlight from technological and economic point of view as well.

Certain types of cationic polymers together with selected inorganics are believed to help to solve this problem. Drilling fluid formulation will depend on type and amount of reactive shales/clays in the actual formations.

Several methods are used to determine shale/clay mineralogy, morphology, concentration, distribution, etc.

Other testing procedures are employed for reactivity evaluation, colloidal characterization, inhibition (prevention of swelling, dispersion), additive's selection and fluid performance evaluation.

An integrated approach promises to achieve the best fluid technology and satisfactory field results.

SHALE INHIBITION AND INHIBITIVE DRILLING FLUIDS

Most of the formations being penetrated during drilling contain certain amount of clay minerals, and so more or less are subject of potential borehole instability. Hydration and swelling of clays lead to causing physical instability (deformation, sloughing, spalling, caving, collapse, etc.) and consequent drilling problems. The composition of shales and their reactivity to water-based drilling fluids can be correlated with cation exchange capacity (CEC). The clay mineral surface is negatively charged and by CEC values shows a good indication of magnitude of possible cationic substitution.

The clay bound water is related to CEC, and this determines the clay's reactivity to fresh water drilling fluids.

The introduction of the potassium based (K-lignite and KCl) drilling fluid in the early '70s generated many discussions and better understanding of clay stabilisation mechanism. K^+ has less hydrational energy than either Na^+ or Ca^{++}. The ionic radius of K^+ in water solution is slightly smaller than the spacing between the clay layers in the montmorillonite matrix. These factors explain the interlayer mobility and specific adsorption of K^+ which is resulted in depressed clay swelling/dispersion, and better borehole stability.

The efforts of drilling mud industry have supported the elimination of disadvantageous Na-ions, by developing the potassium derivatives of conventional organic additives (1).

To increase shale stabilization further, high molecular weight, partially hydrolyzed polyacrylamide has been incorporated into the drilling fluid system. PHPA was used to coat shale sections and encapsulate shale drill cuttings (2). By maintaining a sufficient amount of PHPA polymer, excellent wellbore stability can be achieved (3). The design of shale stabilizing polymer synthesis (4) and special considerations on fluid composition design (5) has led to optimized drilling mud chemistry.

The excess of PHPA by adsorption onto the gumbo shales additionally eliminates bit balling and increases wall cake lubricity (6).

Borehole instability is caused primarily by water transport through shales (7), and is strongly influenced by chemical potentials (8). From this point of view the oil-based muds provide the best performance (10).

New polyglycol (PPG) water-based drilling fluid has been introduced recently as oil-base mud replacement (11).

Cationic organic polymers have been used successfully for clay stabilization in stimulation operations (12).

The translation of this idea into the drilling fluid practice has required extending research and development work, as the conventional additives are essentially anionic in character.

However the strong need for on oil-base mud alternative has led to the development of new cationic polymer drilling fluids (13,14).

These systems have been successfully used in field trials (15,16) and are going to widespread industrial application.

CLAYS RELATED FORMATION DAMAGE PROBLEMS

Most of our sandstone reservoirs contain certain amount of clay minerals finely distributed in the formation matrix. Interlayered, thin shale sections are also frequently present. By using drilling and, or completion fluids of uncontrolled chemistry (without inhibition) serious formation damage has occurred.

The productive formations of our interest are generally fine grained, moderately cemented and not overcompacted. They contain different clay minerals, as illite, kaolinite, chlorite (and in some cases mixed layer clays). Colloid-chemical alteration of authigenic colloidal particles by not properly designed drilling and completion fluid compositions can increase the mobility of fines within the formation. Hydration of the individual crystal lattices that release the colloidal fines reduces the forces necessary to dislodge and suspend the chemically peptized/dispersed particles.

The distribution and potential mobility of clay particles within the formation pore system prompted concern about the possibility of migrating fines causing serious formation damage at high production rate.

It is likely that released clay particles will migrate throughout the entire interval, causing significant formation damage (permeability reduction) especially in low permeability (smaller pore size) formations.

Low damaging drilling/completion fluid design should include several considerations for filtrate chemistry, and filtration mechanism, because the wall cake can retain some of the inhibitive components.

CHARACTERIZATION OF THE SHALES STUDIED

The core and/or cutting samples were analyzed by X-ray diffraction both on powder and an extract of suspension (particles less than 2 μm) to determine the concentration distribution of the clay minerals versus depth. Mineralogy was also evaluated by FTIR (Fourier Transform Infrared Spectroscopy).

Capillary Suction Timer (CST) supported the evaluation of colloidal behaviour in fresh water and electrolyte solutions.

Regarding to formation damage prevention we have used the scanning electron microscopy (SEM) for morphological (structural) description of the available core materials.

EDAX measurements gave important additional informations by elemental analysis. Cation exchange capacity (CEC) was measured by the conventional methylene blue dye adsorption. Additionally a particle charge analyser (PCA) has been used to determine the amount of active anionic chargeable sites being available as adsorptive sites for cationic polymer molecules.

The same method was used to characterize the polymer adsorption under different conditions and to investigate thermal and pH effects.

The critical point from technological point of view is the performance of the selected additives/systems, for that particular shale type. Hot rolling disintegration tests (generally at 90 °C for 16 hours) proved to give the most realistic results and relatively good correlation to field experience.

Statistical analysis of caliper logs of several wells, drilled with different drilling fluid compositions have shown good correlation between percentage of recovered shale cuttings quantity (retained on 2 mm sieve) and prevention of borehole enlargement.

SYSTEM REQUIREMENTS

Different types of drilling fluids such as gyp, lime, K-lignite/KCl, Al-crosslinked polymer, VAMA, PHPA were used for shale drilling with varying degree of success and depending of type (mineralogy) of shales and clay containing formations. So the major task of any developmental drilling fluid is its ability to inhibit water sensitive shales and to minimize hole enlargement. Protection of the environment from polluting chemicals, wastes has become another important requirement. Increasing environmental constraints on the use of diesel-oil containing fluids have led to serious restrictions against (balanced activity) invert-emulsion mud which was successfully used to drill gauge holes. Consequently, there was an urgent need for the drilling fluid industry to develop a water-base mud having inhibitive performance similar to oil muds.

Another major consideration for an inhibitive mud system is the prevention of dispersion of drilled solids. By minimizing dispersion, the integrity of drilled solids can be maintained and effectively removed by mechanical solids control equipment. The high solids removal efficiency will allow the low gravity solids to be maintained below 6 vol% throughout the whole interval to be drilled.

An up-to-date drilling fluid should fulfil the requirements determined by directional and, or horizontal well drilling technology. The application of properly inhibitive mud system helps to maintain borehole stability resulted in gauge hole drilling for better directional control, improved log response and better cement jobs. By its primary function, horizontal well drilling will require efficient formation damage prevention. If the pay zone is contaminated with hydratable/dispersible clays the fluid phase of the drilling (the filtrate) must be as inhibitive as possible, especially for clay particles dispersion/demobilization.

The "time factor" plays very important role either in borehole destabilization or formation damaging. By increasing ROP (rate of penetration) the contact time with the drilling fluid decreases, resulted in reduced risk of borehole instability and, or formation damage. Consequently the selected drilling fluid should enhance ROP together with low impairment and improved lubricity.

EVALUATION OF THE SHALE STABILIZING ADDITIVES

Two shale samples were selected for screening test and detailed evaluation of shale inhibitors.

Kaolinite and montmorillonite were the dominant clay minerals in Shale-A , which was fine grained, lightly cemented and somewhat undercompacted.

Shale-B was composed of mixed layer clay (2%), illite (13%), kaolinite (8%) and chloryte (3%). Other components were quartz, calcite, dolomite, feldspars, etc.

Both formations showed serious borehole instability during drilling with gyp mud. The best known and most widely used inorganic shale stabilizer in the oil industry is the potassium-chloride (KCl). The specific adsorption of potassium ion explains its outstanding role in clays and, or shale stabilization. Benefits of the salts can generally be explained by the collapse of electrical double layer formed by exchangeable cations around the clay layers when ionic strength is increasing.

However the mechanisism of shale stabilization by potassium ion differs from simple ionic effect as it is illustrated in Figure 1. Significantly less KCl is needed to reduce CST (which is in correlation with colloidal stability) to a minimum value, than to achieve reasonable shale(cutting) recovery. This is even more obvious in case of harder, more illitic shale specimen (Figure 2.).

Soluble potassium alone is generally unable to maintain borehole stability and drilled cuttings integrity.

The PHPA polymers are widely used to inhibit troublesome shales, especially when the shale (in bulk) is sensitive to hydraulic effects (erosion).

The selected PHPA polymer provided relatively good shale recovery (Shale-A) in cutting disintegration (hot rolling) tests at very low polymer concentration, as it is shown in Figure 3. Surprisingly the percentage of shale recovery was virtually independent of (PHPA) polymer concentration above the threshold value.

The high molecular weight, cationic polyacrylamide (PAAQ) which was selected regarding to our major task showed a concentration dependent behaviour (see also Figure 3.). It is likely that the different solid-polymer adsorption characteristics and electrostatic interactions are responsible for advanced stabilization achieved by PAAQ addition. This difference was shown to be even more evident in long term disintegration (curing) tests.

Figure 4 shows the results of experiments, where polymers were tested for their ability to inhibit the dispersion of Shale-B.

High efficiency of polymers and consistent stabilizing performance of PAAQ is clearly illustrated.

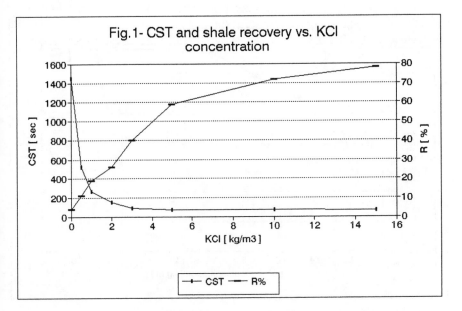

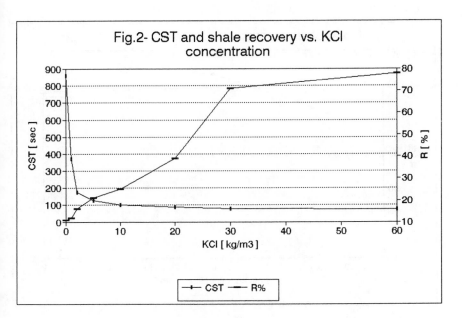

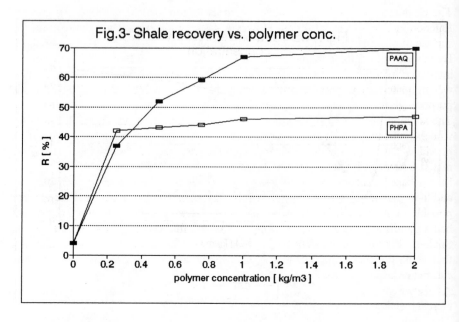

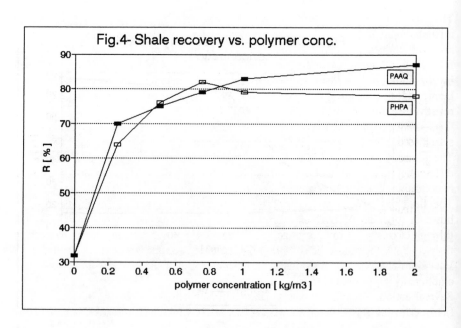

The clear evidence of PAAQ's benefits led us to consider this additive for primary function in borehole stabilization. However the long-chain molecules adsorb on the surface of clay paricles only, but do not penetrate to the interlayer structure of clays. By this way other component(s), having stabilizing effects are needed to prevent the interlayer hydration and swelling. Highly (specifically) adsorbable, low molecular weight quaternary polyamine (QPA), which has better mobility was selected for this purpose.

The addition of QPA to fresh water had a modest impact on dispersion of Shale-A., while significantly better results were gained with Shale-B. At the same time QPA provided excellent additional stabilization in combination with other inhibitive materials, especially in long term tests.

Starch based products are well known in the drilling fluid industry as fluid loss additives. A quaternary modified starch derivative (CMS-2) was evaluated for possible dual function in a cationic polymer system. Figure 5 shows that CMS-2 had a significant impact on dispersion of Shale-A, and even better positive effect on Shale-B. Figure 5 illustrates the result of another interesting interaction, showing that some polymer components can change the adsorption of cationic polymers, causing reduced shale inhibition.

As shown in Figure 6 the addition of CMS-2 to fresh water will substantially increase shale recovery as its concentration is increasing.

The inhibitive effect of the polymers tested is further enhanced by the addition of KCl (or other potassium salts).

FORMULATION OF AN INHIBITIVE CATIONIC DRILLING FLUID

The major problem in formulating a cationic type drilling fluid is the material incompatibility. Most of the additives used in the conventional system are anionic and so incompatible with cationic polymers. Bentonite can be used in prehydrated form, and in limited concentration. However the control of rheology must be solved by the addition of polymeric viscosifiers. Based on extensive laboratory work a hemicellulose base product (HCPB) was selected as primary viscosifier due to its resistivity against chemical and mechanical effects.

The unique chemical structure and performance of welan gum in water-based systems have led to its application as supplementary viscosifier.

The filtration rate control is very important in prevention of formation damage and differential sticking. A starch-base fluid loss additive (SBFLA) proved to be effective in controlling the filtration rate of cationic drilling fluid.

The fluid pH is necessary to control from a corrosion inhibition point of view; however too high pH will cause increased clay dispersion as it was clearly shown through lab tests.

Consequently the pH of the cationic drilling fluid should be kept near to neutral.

Potassium-chloride will improve the inhibitive characteristic of the cationic mud, and additionally suppresses chemical's incompatibility.

Several cationic fluid compositions were investigated for shale inhibition, stability, field performance, potential damaging, etc. Based on these results and experience

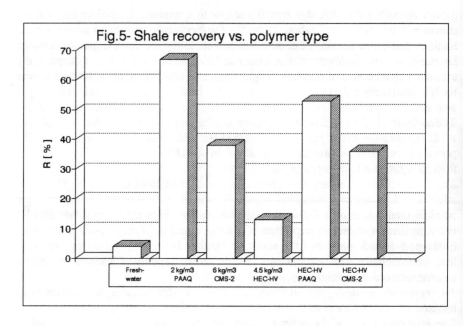

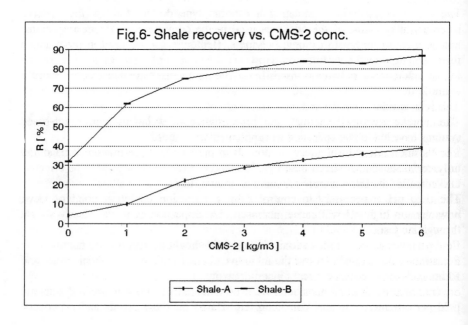

gained through field trials the typical composition advised for field application is shown in Table 1.

Shale-A due to its slightly cemented and undercompacted character is sensitive to hydraulic effects. In Figure 7 the effect of fluid viscosity (bentonite suspension), cationic polymer, and fluid composition (cationic polymer mud without and with KCl) is illustrated.

The results formulated the guideline for field application of the cationic polymer drilling fluid.

FIELD APPLICATION OF CATIONIC POLYMER DRILLING FLUIDS

At the beginning cationic drilling mud has been used in vertical wells to penetrate shale formations which have suffered serious borehole enlargement when were drilled with fresh-water and, or gyp mud.

The improvement was necessarily needed because the worst of the borehole enlargement was occurring near the productive formation.

The conventional drilling muds did not inhibit the shale, and excessive sloughing resulted.

Composition of the cationic system was almost the same as it is shown in Table-1. KCl was added to improve the compatibility of the additives and to enhance shale inhibition.

It vas very important to maintain the salt concentration by checking the chlorides concentration several times a day (K-ion occasionally) and by adding any needed salt.

Rheological properties were maintained at relatively high range of values at upper hole section where hole erosion was assumed to be caused by clay dispersion rather than swelling.

Drilling fluid properties were reproduced in the field better than expected prior to the operations.

This is shown in Table 2 through comparison of lab and field fluid parameters.

The improvements to the actual mud program were based on lab reactivity tests on cuttings from the offset wells and from the well being drilled.

Results summarized in Figure 7 clearly show the excellent performance of cationic polymer muds (CPM-1 and CPM-2).

Drilled solid tolerance of polymer muds is known to be very modest. Solids control efficiency was monitored on a frequent basis; the equipment could be fine tuned with changes in lithology, thus maintaining optimum efficiency.

Low gravity solids should be kept to a minimum. The range for optimum drilling fluid performance was to limit the low gravity solids to less than 6 vol% and maintain a maximum CEC value of 40 kg/m3.

Excessive solid content can cause dramatic increase in rheology. However conventional dispersants must not be used in a cationic system. Special nonionic polymer was advised to use, but proved to be practically uneffective.

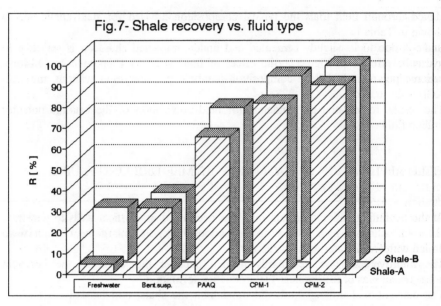

Table 1 - TYPICAL COMPOSITION OF CATIONIC POLYMER FLUID

Bentonite	5 - 15	kg/m^3
HCBP	5 - 10	kg/m^3
PAAQ	1.5 - 2.0	kg/m^3
QPA	3.0 - 4.5	kg/m^3
Welan-gum	1.0 - 2.0	kg/m^3
SBFLA	12 - 18	kg/m^3
KOH	0.5 - 1.0	kg/m^3
KCl	35 - 70	kg/m^3

TABLE 2 - PROPERTIES OF LAB AND FIELD CATIONIC POLYMER MUDS

Fluid composition		Lab -1	Field - 1	Field - 2	Field - 3
Density (kg/m3)		1060	1150	1130	1140
Fann readings	600	39	41	35	38
	300	28	28	23	26
	200	22	20	17	20
	100	17	13	12	15
	6	7	3	3	4
	3	5	2	2	3
Gel strength 10" (Pa)		2.1	1.5	1.5	1.5
Gel strength 10' (Pa)		5.1	5.1	4.1	5.6
Filtration - API (ml)		8.4	5.8	6.2	5.3
Filtration - HPHT (ml)		36.8	19.9	18.5	17.2
pH (-)		8.42	8.23	8.58	8.06
Cl - ion conc. (g/l)		31.2	40.4	31.9	33.6
Clay content (kg/m3)		14.2	25.0	21.3	21.3
HPHT filtration was measured at 110 C° - 3.5 MPa					

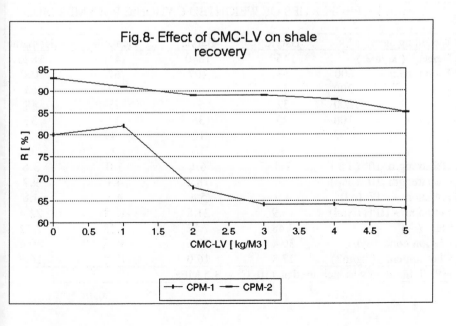

Fig.8- Effect of CMC-LV on shale recovery

We have done several screening tests and surprisingly the low-viscosity CMC showed the best performance in eliminating solids caused rheology problems. The open question was how CMC-LV will influence on shale inhibition. Results in Figure 8 show certain reduction of shale recovery in absence of KCl, but at higher CMC-LV concentrations than is generally used.

Caliper logs in the wells drilled with cationic polymer muds show gauge hole for the whole intervals.

In directional and horizontal wells the field results were even more impressive. Several hundred meters were drilled trouble-free in horizontal well sections.

An additional advantage of this approach was a reduction in environmental impact due to the dumping of minimum waste drilling fluid.

In case of high pressure drilling the cationic polymer mud can be weighted using conventional weighting materials. Results summarized in Table 3. show fluid parameters of lab weighted field mud after hot rolling at 110 °C for 16 hours. Possible success of weighted cationic polymer system application will however particularly depend on solids control efficiency

TABLE 3 - PROPERTIES OF WEIGHTED CATIONIC POLYMER MUDS				
Weighting agent	Baryte	Baryte	Hematite	Hematite
Density (kg/m3)	1450	1840	1410	1820
Fann readings 600	68	107	67	100
300	42	68	37	63
200	32	51	28	48
100	20	36	17	35
6	4	12	3	7
3	3	10	2	6
Gel strength 10" (Pa)	1.0	6.1	1.0	3.6
Gel strength 10' (Pa)	5.6	18.4	4.1	9.7
Filtration - API (ml)	4.9	5.4	4.5	4.8
Filtration - HPHT (ml)	18.9	21.8	19.8	22.4
pH (-)	7.48	7.57	7.58	7.72
Cl - ion conc. (g/l)	30.4	30.6	30.6	30.6
Clay content (kg/m3)	17.8	16.0	17.8	16.0
HPHT filtration was measured at 110 C° - 3.5 MPa				

CONCLUSIONS

Based on the lab and field results the following conclusions have been made:

1.)   Cationic polymer systems (CMS) should be formulated for maximum shale inhibition considering additive's compatibility.

2.)   CMS can be used for long term shale stabilization, and trouble-free shale drilling.

3.)   Solid (and clay) content should be kept to minimum by using polymeric viscosifiers rather than bentonite.

4.)   Rheology problems caused by drilled solids accumulation - within practical limits - can be controlled by addition of CML-LV (other anionic materials, are also under evaluation).

5.)   CMS represents a low impairment drilling fluid with controlled filtrate chemistry for drilling trough productive zones in horizontal wells.

6.)   Reuse of CMS reduces costs and fluid disposal volumes.

<u>References</u>

1.)   Palumbo, S., Giacca, D., Ferrari, M. and Pirovano, P.: The development of potassium cellulosic polymers and their contribution to the inhibition of hydratable clays. SPE 18477 (1989) SPE Int. Symp. on Oilfield Chemistry, Houston.

2.)   Clark, R.K. et al.: Polyacrylamide/potassium-chloride mud for drilling water-sensitive shales. Journal of Petroleum Technology (June 1976) 719-726.

3.)   Kadaster, A.G., Guild, G.J., Hanni, G.L. and Schmidt, D.D.: Field application of PHPA muds. SPE Drilling Engineering (Sept. 1992) 7.191-199.

4.)   Sheu, J.J. and Perricone, A.C.: Design and synthesis of shale stabilizing polymers for water-based drilling fluids. SPE 18033 (1988) SPE Annual Technical Confereence and Exhibition, Houston.

5.)   Chesser, B.G.: Design considerations for inhibitive, stable water-based mud systems. SPE 19757 (1989) IADS/SPE Drilling Conference, Dallas.

6.)   Roy,S. and Cooper, G.A.: Prevention of bit balling in shales - Preliminary results. SPE Drilling and Completion (Sept. 1993) 8. 195-200.

7.)   Ballard, T.J., Beare, S.P. and Lawless, T.A.: Fundamentals of shale stabilization: Water transport through shales SPE 24974 (1991) European Petroleum Conference, Cannes.

8.)   Hale, A.H., Mody, F.K. and Salisbury, D.P.: The influence of chemical potential on wellbore stability. SPE Drilling and Completion (Sept. 1991) 8. 207-216.

9.)   Beihoffer, T.W., Dorrough, D.S., Deem, C. K., Schmidt, D.D. and Bray, R.P.: Cationic polymer drilling fluid can sometimes replace oil-based mud. Oil and Gas J. (Mar.16, 1992) 47-52.

10.) Bol, G.M., Sau-Wai Wong, Davidson, C.J. and Woodland, D.C.: Borehole   stability in shales SPE 24975 (1991) European Petroleum   Conference,   (France)

11.) Enright, D.P., Dye, W.M. and Smith, F.M.: An environmentally safe water-based alternative to oil muds. SPE Drilling Engineering (March 1991) 7.15-19.

12.) Hild, D.G.: Clay stabilization - Criteria for best performance. SPE 10656 (1982) SPE Formation Damage Control Symp., Lafayette.

13.) Beihoffer, T.W., Dorrough, D.S., Deem, C.K., Schmidt, D.D. and Bray, R.P.:Cationic polymer drilling fluid can sometimes replace oil-based mud. Oil and   Gas J. (Mar.16, 1992) 47-52.

14.) Retz, R.H., Friedheim, J., Lee, L.J. and Welch, O.O.: An environmetally acceptable and field practical, cationic polymer mud system. SPE 23064 (1991) Offshore Europe Conference (Aberdeen).

15.) Hemphill, T., Valenziano, R., Bale, P., and Sketchler, B.: Cationic drilling fluid improves ROP in reactive formations.Oil and Gas J.(June 8, 1992) 60-65.

16.) Welch, O., and Li-Jein Lee: Cationic polymer mud solves gumbo problems in North Sea. Oil and Gas J. (July 13, 1992).

Chemistry and Function of Chromium in Lignosulfonate and Lignite Thinners. Development of Environmentally-Friendly Aqueous Drilling Fluids

F. Miano, S. Carminati, and T. P. Lockhart

ENIRICERCHE SpA, 20097 SAN DONATO MILANESE, ITALY

G. Burrafato

AGIP SpA, 20097 SAN DONATO MILANESE, ITALY

ABSTRACT
New hypotheses regarding the roles of Cr(VI) and Cr(III) in determining the thinning power of Cr-lignosulfonates and Cr-lignites and in stabilizing mud rheology at elevated temperature have been formulated on the basis of experiments on bentonite suspensions, combined with an analysis of some published results. These ideas have led to the evaluation of Cr(III) and Zr complexes as additives in chrome and chrome-free bentonite muds formulated with lignosulfonate and lignite thinners. The results demonstrate the functional equivalence of Cr(III) complexes to Cr(VI), and show that bentonite muds formulated with the Zr complexes and chrome-free lignosulfonate and lignite thinners exceed traditional chrome muds in their stability to aging at elevated temperature.

INTRODUCTION
Large volumes of drilling fluids are employed in the drilling of petroleum and gas wells. The common practice of disposing drill cuttings at the well-site brings at least a part of these fluids into contact with the environment.[1] Studies showing that traditional oil-based fluids are toxic to marine organisms have led to progressively stricter regulations on the use of oil muds, particularly off-shore. These environmental constraints have stimulated considerable effort to identify less toxic drilling fluids that preserve, as much as possible, the beneficial performance characteristics of the oil-based muds.[1-4]

In this context there has been strongly renewed interest in the development of aqueous drilling fluids. Aqueous fluids, however, have significant performance limitations that must be resolved in order for them to constitute viable and economic alternatives to oil-based muds over the full range of drilling situations. In particular, present aqueous fluids have: (1) limited applicability at high temperature, especially when they contain high levels of drill solids; (2) higher reactivity toward shales than oil-based muds, resulting in increased borehole instability and other operational problems; (3) an increased tendency, relative to oil-based muds, to damage the producing formation. A fourth consideration must be added to the above: Notwithstanding the low toxicity of the base fluids (water, bentonite clay, polymer), some of the other common constituents are in fact toxic and are increasingly coming under regulation. Thus, in addition to performance and economic issues, aqueous drilling muds must also resolve toxicity problems of their own.[1-3]

Our program to develop improved aqueous drilling fluids has focused on aqueous bentonite (high solids) muds. The rheological properties of these fluids are achieved through the use of bentonite clay (typically 3-6% by weight) together with chrome or ferrochrome (Cr- or FeCr-)lignosulfonate and chrome lignite (Cr-lignite) thinners, the latter being preferred at elevated temperatures and having a second function of controlling fluid loss.[4] Our interest in these bentonite muds stems from the fact that they are the least expensive and most widely-employed drilling fluids in use today, and that there is considerable field experience with their application to a variety of drilling conditions.

We have focused our attention, in particular, on what we perceive to be two key limitations of these fluids: (1) their limited resistance to high temperatures, which excludes their being used for the drilling of deep, hot formations without extensive dilution, high rates of thinner maintenance, and active removal of drill solids,[5] and (2) the presence of toxic Cr(VI) ions which are added in order to boost the resistance of the lignosulfonate and lignite thinners at elevated temperature.[6] The presence of Cr(VI) and, to a lesser degree Cr(III), in these drilling muds raises increasingly pressing questions about their long-term viability as economic alternatives to other aqueous and oil-based muds.

Previous efforts to resolve the Cr(VI) problem in bentonite muds have followed various strategies. For example, chemical reaction of Cr(VI) with the thinner followed by chemical separation has been employed in order to obtain a Cr-lignosulfonate and Cr-lignite in which chromium is present predominantly as the Cr(III) ion. Another approach has been to synthesize lignosulfonates complexed with metal ions (*eg.*, Fe and Ti) different than chromium.[1,3] Although the thinners obtained with these approaches do not provide high temperature resistance or solids dispersing capacity equivalent to Cr(VI)-containing mud formulations, several have been commercialized.

In this paper we will describe a novel approach articulated on the basis of our current ideas on the chemical role of Cr(VI) in stabilizing bentonite muds, and we will demonstrate that it is possible to formulate Cr(VI)-free aqueous bentonite muds possessing excellent resistance to elevated temperature.

EXPERIMENTAL
Materials.
The lignosulfonates and lignites employed in this study were obtained from commercial sources. Chemical analyses (by atomic adsorption or plasma activation) on the chrome-thinners as supplied indicated chromium levels ranging from 1.2% to 2.6% by weight. A colorimetric analysis on the commercial "chromate-free" Cr-lignite employed showed it to contain an estimated 0.1% of Cr(VI). The Cr(III) complexes and Zr-citrate were prepared as described in refs.7 and 8. Wyoming bentonite (API specification grade) was used in all of the mud formulations. In some drilling fluid formulations an Italian outcrop shale was employed in order to simulate mud contamination by active solids. This shale contains 50% smectites, and was added to the formulations as a fine powder (95% of particles <16 μm). Commercial barite (API specification grade) was employed as a weighting agent.

Procedures.
The drilling fluids described in the text were prepared with dispersions of bentonite clay hydrated for at least 16 hours to which the other additives were added. The suspensions were mixed vigorously (Hamilton Beach mixer) after the addition of each component and the mud pH was adjusted with a 25% NaOH solution. The muds were aged in pre-heated ovens at elevated temperature under dynamic conditions (hot rolling at 17 rpm). Before measurement of mud rheology, heat-aged samples were mixed at low speed for 10 min with a Hamilton Beach mixer.

Three different rheometers were employed in the course of this study. The mud rheologies reported in Tables II and IV were measured at ambient temperature on a Fann Model 35 rheometer and the plastic viscosity, PV, apparent viscosity, AV, yield point, YP, and the 10" and 10' gel strengths were derived according to the API procedure.[9] Table I reports mud rheologies (AV, PV, and YP) derived from measurements on a Bohlin CS rheometer at two shear rates (200 s^{-1} and 400 s^{-1}). The rheologies reported in Table III were obtained on a Bohlin VOR rheometer. The PV values were derived from measurements at 500 and 1000 s^{-1}, while the YP was obtained by extrapolating to zero shear from the data measured between 1 and 10 s^{-1}. The elastic modulus, G', was measured operating the rheometer in the oscillatory mode after waiting 10 minutes for the sample to equilibrate.

Dialysis of a commercial Cr-lignite. A 10 wt % aqueous solution of Cr-lignite prepared at pH 10 was divided into two portions which were placed into dialysis membranes. The lignite solutions were then dialyzed against a large volume of aqueous solution containing either 0.1 N sodium acetate or sodium oxalate. After allowing two weeks for equilibration, the external solution was a transparent golden color, indicating little loss of the lignite through the membrane. The lignite in the dialysis bag was recovered, dried to a solid, and then used to formulate the drilling fluids described in the text.

RESULTS AND DISCUSSION
Chemical forms of chromium in lignosulfonate and lignite thinners.
The technical literature provides relatively few indications as to the forms of chromium [Cr(VI) and Cr(III)] present in lignosulfonates and lignites and, in particular, of the role that chromium plays in determining the rheological properties of bentonite suspensions. In a key paper on the synthesis of Cr-lignosulfonate, James and Tice[10] found that Cr(VI) oxidation of lignosulfonate leads to a 20-fold increase in its molecular weight, which they postulated to occur via phenolic coupling. They also found considerable uptake by the organic matrix of the Cr(III) byproduct, which they assumed to be bound to carboxylate groups also formed during the oxidation. In fact, by eluting the as-formed Cr-lignosulfonate gel with acetic acid or HCl solution, some 87% of the chromium was recovered without major change to the product. Digesting the Cr-lignosulfonate gel with a strong binding agent (EDTA) led to the removal of the remaining 13% of chromium, the loss of which caused the aqueous gel to revert to a fluid solution. Their interpretation of these results was that most of the chromium is present as Cr(III) and is either unbound, or weakly bound, to the organic matrix, while a smaller portion of Cr(III) is strongly-bound and acts as a crosslinking agent for the aqueous gel.

More recently, Pettersen[11] carried out chromatographic studies on a commercial Cr-lignosulfonate and Cr-lignite [both of which had been prepared by oxidative reaction with Cr(VI)] and found that a significant fraction of the chromium in the thinners was in fact present as Cr(VI). While these two studies provide useful information on the chemical forms of chromium present in Cr-lignosulfonates and the closely related Cr-lignites, they do not offer insight as to whether or how these chromium species influence the properties of these thinners.

The function of Cr(III).

In recent papers[12,13] we have probed the nature of the surface interaction of synthetic and lignosulfonate thinners with bentonite clay in water. These studies showed that, whereas a lignosulfonate in the sodium form interacted very little with the surface of bentonite particles, FeCr-lignosulfonate binds much more strongly, and is a much more effective thinner. Bridging of Cr(III) and Fe(III) between the lignosulfonate and the clay surface was invoked in order to account for the enhanced lignosulfonate adsorption.

We report here new experiments, carried out on lignites, that provide further support for the key role played by Cr(III) in enhancing the thinning ability of lignosulfonate and lignite thinners. These experiments are based on a dialysis procedure (see *Experimental*) by means of which chromium [both Cr(VI) and Cr(III)] is removed from a commercial Cr-lignite. By employing progressively stronger Cr(III) complexing agents in the dialysis medium it is possible, by analogy with the chemical digestion reported by James and Tice,[10] to extract progressively larger portions of the chemically-bound Cr(III) originally present in the Cr-lignite. Analysis of the dispersing power of the dialyzed lignite therefore provides insight into the role that the extracted chromium played in determining the thinning power of the original Cr-lignite. Further, comparison of the rheological properties of the bentonite suspensions after aging at elevated temperature provides an indication of the importance of the extracted chromium in determining the resistance of the suspension to aging.

Results of such experiments are summarized in Table I, which reports the rheological properties of a series of bentonite suspensions containing Na-, Cr-, and dialyzed Cr-lignites before and after aging at 150°C. A comparison of the initial rheological properties of the dispersions shows clearly the enhanced thinning ability of the original Cr-lignite (entry 2) relative to the acetate- and oxalate-extracted Cr-lignites (entries 3 and 4), which have AV and PV values intermediate between those of the original Cr-lignite and the lignite-free suspensions. The removal of chromium from the original Cr-lignite thus has clearly compromised its thinning ability.

After aging of the bentonite suspensions, the lignite-free and Na-lignite-thinned suspension displayed similar (high) rheological values relative to the suspension containing the commercial Cr-lignite thinner. Once again, the acetate and oxalate-extracted lignites displayed intermediate behavior, indicating that the chromium extracted (which included the small amount of Cr(VI) present in the sample) plays a role in stabilizing the suspension.

Cr(VI) stabilization of lignosulfonate and lignite thinners.

Having ascertained that the presence of Cr(III) is important for the thinning activity of lignite and lignosulfonate, it is important to clarify the functional role of Cr(VI), particularly inasmuch as this is the more toxic and strictly regulated form of chromium. It has long been known[6,14] that the addition of Cr(VI) (as sodium chromate or dichromate) to lignosulfonate- and lignite-thinned drilling fluids greatly enhances their stability at elevated temperature, and even allows the rheological properties of exhausted muds to be recovered, to some degree. Still today, in fact, many lignites are furnished as simple mixtures with chromate or dichromate. Thus, mud maintenance at elevated temperature by lignite addition[2,6] also replenishes the Cr(VI). As Pettersen showed,[11] however, even so-called Cr(VI)-free Cr-lignosulfonates or Cr-lignites, prepared by reaction of dichromate with lignite, may contain significant residual Cr(VI).

Table I. Influence of lignite on the rheology of bentonite suspensions at 25°C before and after aging at 150°C for 18h.

Entry	Type of lignite	Lignite dialyzed against	Rheology (before/after aging)		
			AV (mPa*s)	PV (mPa*s)	YP (Pa)
(1)	none	--	41/118	23/37	18/81
(2)	Cr	--	25/66	11/41	14/25
(3)	Cr	acetate	27/97	13/57	14/40
(4)	Cr	oxalate	33/81	17/55	16/26
(5)	Na	--	27/119	13/53	14/66

Composition: Bentonite clay, 7%; lignite, 1%; initial pH 10-12.

We have carried out several experiments in order to further elucidate the impact of Cr(VI) on the properties and thermal resistance of bentonite muds. Table II reports the rheological properties of a series of clay dispersions thinned with either Na-lignosulfonate or a commercial (chrome-free) Fe-lignosulfonate with and without added Cr(VI) (furnished as potassium dichromate). Comparison of the initial rheological properties of the Na- and Fe-lignosulfonate suspensions establishes that the presence of Cr(VI) does not enhance the thinning ability of the lignosulfonates. On the other hand, focusing attention on the YP and the gel strengths at 10" and 10', which are the parameters most sensitive to the strength of the interaction between the dispersed bentonite particles, we find that the presence of Cr(VI) stabilizes (or, in the case of the Na-lignosulfonate suspensions actually improves) the rheology of both the Na- and Fe-lignosulfonate suspensions to aging at 180°C. In summary, *Cr(VI) is not itself a thinner or a thinning aid, but it has a pronounced ability to maintain or even improve the rheology of bentonite suspensions during aging at elevated temperature.*

Hypotheses on the mechanism of Cr(VI) stabilization of bentonite drilling fluids.

The beneficial effect of Cr(VI) on mud stability has been associated[14] with its oxidation of the lignite and lignosulfonate thinners present in the drilling fluid. In light of the importance of Cr(III) to the thinning-ability of

Table II. Bentonite clay dispersions prepared with Cr(VI) and Na- or FeCr-lignosulfonates before and after aging 16 h at 180°C.

Composition		Rheology (before/after aging)				
Form of Ligno-sulfonate	Wt% Cr(VI)	AV (mPa*s)	PV (mPa*s)	YP (Pa)	Gel Strength (Pa)	
					10"	10'
Na	--	31/36	14/22	17/14	20/7	37/18
Na	0.07	35/43	15/34	20/9	24/2	35/9
Na	0.2	42/52	18/45	24/7	29/1	29/1.5
Fe	--	19/33	17/27	2/6	0.5/1.5	1/13
Fe	0.07	19/21	17/20	2/1	0.5/1	1/1.5

Composition: Bentonite clay, 7%; lignosulfonate, 1%; Cr(VI) added as $K_2Cr_2O_7$; initial pH 10.

lignosulfonate and lignite, we propose that the oxidation of the lignite/lignosulfonate organic matrix by Cr(VI) is important to the rejuvenation or maintenance of bentonite muds principally because it furnishes additional Cr(III) to the thinner.

This hypothesis implies that the functional form of Cr(III) initially present in a Cr-lignite- or Cr-lignosulfonate-thinned mud is lost during extended exposure to elevated temperature. In fact, it is well-known that Cr(III) is strongly driven at alkaline pH toward hydrolysis and precipitation as the extremely insoluble hydroxide, $Cr(OH)_3(H_2O)_3$.[15] We presume, therefore, that Cr(III) is converted from its active form (bound to the thinner) to the stable and *inactive* form, $Cr(OH)_3(H_2O)_3$, during aging of the drilling fluid.

In contrast to Cr(III), Cr(VI) is soluble even at high pH. In view of these considerations, we propose the following explanation for the stabilizing/rejuvenating influence of Cr(VI) on bentonite dispersions thinned with lignosulfonate or lignite: *Cr(VI) functions as a soluble precursor of Cr(III), which it furnishes to the drilling fluid at a controlled rate at elevated temperature as a consequence of its reaction with the organic matrix. In this way, the thinner is compensated for Cr(III) lost as $Cr(OH)_3(H_2O)_3$ through base hydrolysis.*

Temperature-resistant Cr(VI)- and Cr(III)-free lignosulfonate/lignite muds.

These ideas as to the roles played by Cr(III) and Cr(VI) in determining the performance of bentonite muds thinned with Cr-lignosulfonate and Cr-lignite provided new bases for exploring ways to achieve good mud stability even in the absence of Cr(VI) [or Cr(III)]. The particular challenge that we posed ourselves was that of identifying other chemical additives capable of duplicating the functional roles played by Cr(VI) [and, if possible, Cr(III)].

In approaching this problem we focused first on the presumed role of Cr(VI) as an alkaline pH-stable reservoir of Cr(III) for the thinner. We were aware that certain complexed forms of Cr(III) possess considerably better kinetic and thermodynamic stability with respect to $Cr(OH)_3(H_2O)_3$ formation than do simple inorganic salts of Cr(III).[7,15] Hence we felt that it might be possible to identify one or more Cr(III) complexes capable of furnishing Cr(III) to the lignosulfonate or lignite thinner in a controlled way at elevated temperature.

To this end, a number of Cr(III) compounds were evaluated for their ability to stabilize lignosulfonate- and lignite-thinned bentonite dispersions. Whereas simple inorganic Cr(III) salts (*eg.*, $CrCl_3$) and chromium complexes of low stability [*eg.*, $Cr(acetate)_3$] proved to be largely or wholly ineffective, other complexes displayed a range of mud-stabilizing abilities. Table III summarizes results for bentonite muds containing a high level of a reactive shale contaminant and thinned with FeCr-lignosulfonate and either Na- or Cr-lignite. The Na-lignite muds formulated with $K_2Cr(glycolate)_3$ and $K_3Cr(oxalate)_3$ possessed markedly better rheological properties after aging at 180°C than the mud containing Na-lignite alone. Remarkably, the Na-lignite muds formulated with the Cr(III) complexes achieved better control over the YP and elastic modulus than even the traditional Cr-lignite mud included for comparison.

An interesting and direct comparison on Na-lignosulfonate/bentonite suspensions (Table IV) shows that $K_3Cr(oxalate)_3$ and Cr(VI) have a remarkably similar impact on the evolution of the rheology during heat aging. This result strongly supports our affirmation above that the oxidation of the lignosulfonate or lignite matrix, of itself, is not the key to interpreting the

Table III. Rheology of FeCr-lignosulfonate/lignite muds containing bentonite, reactive shale and Cr(III) complexes before and after aging at 180°C for 16 h.

Composition		Rheology (initial/after aging)		
Type of lignite	Cr(III) complex	PV (mPa*s)	YP (Pa)	Elastic modulus, G' (Pa)
--	--	23/24	0.4/6.5	3.1/444
Na	--	22/32	0.5/1.4	3.7/82
Na	$K_3Cr(oxalate)_3$	21/28	0.3/0.4	3.2/15
Na	$K_2Cr(glycolate)_3$	21/16	0.6/0.7	3.7/17
Cr	--	22/16	0.7/1.8	3.0/95

Composition: Bentonite, 7%; FeCr-lignosulfonate, 1.5%; lignite, 0.5%; Cr(III) complex, 0.2%; reactive shale, 15%; initial pH 10.

rheology-stabilizing influence of Cr(VI), but that it is the generation of Cr(III) that is important.

The above results, and numerous others on realistic drilling mud formulations have established the ability of these Cr(III) complexes to act as effective high temperature extenders for bentonite drilling fluids formulated with commercial thinners (both chrome and chrome-free). To the degree that this approach provides a valid means for eliminating Cr(VI) without sacrificing mud stability at elevated temperature, we feel that these results constitute a significant realization of our first objective.

We have also tackled the more ambitious problem of finding chrome-free additives capable of substituting for the chemical functions of both Cr(VI) and Cr(III). Particularly exciting results have been obtained with certain organic complexes of zirconium. Figure 1 shows the influence of Zr-citrate on the stability of bentonite drilling fluids formulated with commercial chrome-free Fe-lignosulfonate and lignite thinners. The resistance of the chrome-free mud formulations to extended aging at elevated temperature (400 hours/150°C) is markedly enhanced in the presence of the zirconium additive. The Zr-citrate containing, chrome-free mud, in fact, even outperformed the traditional chrome mud reference significantly; we consider this result exceptional in light of previous experiences.[1,3]

Table IV. Comparison of the relative rheology-stabilizing capacity of Cr(VI) and $K_3Cr(oxalate)_3$ in Na-lignosulfonate bentonite muds aged at 120°C for 16 h.

	Rheology (initial/after aging)				
Chromium	AV (mPa*s)	PV (mPa*s)	YP (Pa)	Gel Strength (Pa) 10"	10'
Cr(VI)	38/44	13/37	25/7	27/1	26/1.5
$K_3Cr(oxalate)_3$	30/22	14/19	16/3	20/2.5	39/7

Compositions: Bentonite, 7%; Na-lignosulfonate, 1%; Cr(VI), 0.07% (as $K_2Cr_2O_7$); Cr(III), 0.07% as oxalate complex; initial pH 10.

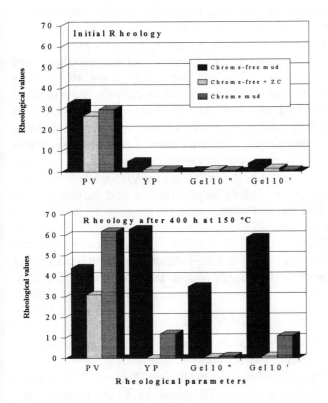

Figure 1. Influence of Zr-citrate on the rheology of bentonite muds before and after aging at 150°C for 400 h. Compositions: For all muds, bentonite (5.4%), chrome-free lignite (1.2%), barite (to density 1.8 Kg/L), initial pH 10; for the chrome-free muds: Fe-lignosulfonate (1.2%), Zr-citrate (0 or 0.4%); for the chrome muds: FeCr-lignosulfonate (1.2%); Cr-lignite (0.4%).

We have carried out extensive testing of the zirconium additives in complex mud formulations at temperatures as high as 200°C and to mud densities exceeding 2.0 kg/L. These tests show that the new additives are effective both as high temperature *extenders* for chrome muds and for the formulation of chrome-free muds possessing excellent high temperature stability. Not least among the merits of the zirconium additives is that zirconium is economically viable as a substitute for chromium and possesses low toxicity.

CONCLUSIONS
We have pursued the development of Cr(VI)-free aqueous bentonite muds possessing the following characteristics: (1) compositions as similar as possible to those of the traditional Cr-lignosulfonate/lignite muds in order to facilitate their introduction into field operations, (2) cost comparable to the current chrome muds, and (3) good resistance to elevated temperature.

New and published results have been used to develop new hypotheses as to the function of Cr(VI) and Cr(III) in Cr-lignosulfonate and Cr-lignite thinners. These, in turn, have provided a rewarding basis for pursuing new avenues toward the preparation of more thermally-stable and environmentally-friendly bentonite muds. We have shown that certain Cr(III) complexes are capable of duplicating the chemical role of Cr(VI), and offer the possibility of achieving the current level of performance of Cr(VI)-containing lignosulfonate/lignite muds without employing this toxic and strictly-regulated chemical. It has also been found that zirconium complexes, which have the attributes of moderate cost and low toxicity, are extremely effective substitutes for both Cr(III) and Cr(VI): muds formulated with Zr-citrate and commercial Cr-free lignosulfonates and lignites display thermal stability exceeding that of chrome muds.

ACKNOWLEDGMENTS
The authors thank D. Giacca (Agip) and L. Faggian (Eniricerche) for their contributions to this work and Agip SpA and the ENI Group for their generous support.

REFERENCES
1. Bleier, R.; Leuterman, A.J.J.; Stark, C. *J. Petrol. Techn.* (1993) *45*, 6.
2. Plank, J.P. *Oil Gas J.* (Mar. 2, 1992) 40.
3. Park, L.S. "A New Chrome-Free Lignosulfonate Thinner: Performance Without Environmental Concerns," SPE paper 16281 presented at the SPE Intl. Symp. on Oilfield Chem., San Antonio, Feb. 4-6, 1987.
4. Lundie, P.R. "Drilling Fluids: A Challenge to the Chemist," Proc. 3rd Intl. Symp. on Chemicals in the Oil Industry, Univ. of Manchester, April 19-20, 1988.
5. Thurber, N.E. "Waste Minimization for Land-Based Drilling Operations," SPE paper 23375 presented at the 1st Intl. Conf. on Health, Safety and Evironment, The Hague, Nov. 10-14, 1991.
6. Kelly, J. *Oil Gas J.* (Oct. 5, 1964) 112.
7. Albonico, P. Burrafato, G.; Lockhart, T.P. "Effective Gelation-Delaying Additives for Cr^{+3}/Polymer Gels," SPE paper 25221 presented at the SPE Intl. Symp. on Oilfield Chemistry, New Orleans, (March 2-4, 1993).
8. Ermakov, A.N.; Marov, I.N.; Kazanskii, L.P. *Russ. J. Inorg. Chem.* (1967) *12*, 1437.
9. "Recommended Practice Standard Procedure for Field Testing Water-Based Drilling Fluids," RP 13B-1, American Petroleum Institute, 1990.
10. James, A.N.; Tice, P.A. *Tappi*, (1964) *47*, 43.

11. Pettersen, J.M. *Anal. Chim. Acta* (1984) *160*, 263.
12. Rabaioli, M.R.; Miano, F.; Lockhart, T.P.; Burrafato, G.: "Physical/Chemical Studies on the Surface Interactions of Bentonite with Polymeric Dispersing Agents," SPE paper 25179 presented at the SPE Intl. Symp. on Oilfield Chem., New Orleans, March 2-5, 1993.
13. Rabaioli, M.R.; Miano, F. *Colloids and Surfaces*, in press.
14. Skelly, W.G.; Dieball, D.E. *Soc. Petrol. Eng. J.* (June, 1970) 140.
15. Hartford, W.H.; Copson, R.L "Chromium Compounds," in Kirk-Othmer Encylopedia of Chemical Technology, 3rd Ed., Wiley, New York (1979), vol. 6 pp.88,89.

Mixed Metal Hydroxide (MMH) – A Novel and Unique Inorganic Viscosifier for Drilling Fluids

J. Felixberger

SKW TROSTBERG AG, DR. ALBERT-FRANK-STRASSE 32, PO BOX 1262, D-83308 TROSTBERG, GERMANY

1. Introduction

Many requirements are imposed on the drilling fluid system, e. g., controlling the subsurface pressure, cooling and lubricating the bit, forming a thin, elastic, low-permeability filter cake on walls of the borehole, avoiding permeability damage to producing formation, stabilizing the borehole, being environmentally benign etc.

But the major function of the drilling fluid is to transport cuttings from beneath the bit, carry them up the annulus as quickly and efficiently as possible and support their removal at the surface.

Good hole cleaning minimizes drilling problems such as bit balling, drill string torque and drag, borehole instability, slow drilling rate etc.

A suitable drilling fluid rheology is critical to successful drilling operations. For example, horizontal drilling and milling operations require a highly shear-thinning fluid for maximum drilling efficiency.

Viscosifiers, such as bentonite and polymers are used to obtain the right fluid rheology and flow profile.

2. Drilling Fluids Rheology [1,2]

Drill cuttings slip through the drilling fluid due to gravity. The upward velocity of the drilling fluid and its buoancy counteract the gravitational force.

Consequently, the resulting velocity of the cuttings v_r is the difference between the fluid velocity v_{fl} and the slip velocity of the cuttings v_s.[1]

$$v_r = v_{fl} - v_s$$

From this it follows that increasing annular fluid velocity increases the upward velocity of the cuttings too. There is another phenomenon which supports the lifting of cuttings with increasing fluid velocity. At a critical velocity there is a transition from laminar flow regime to turbulent flow regime (Fig. 1).

Fig. 1: Laminar and turbulent flow regime

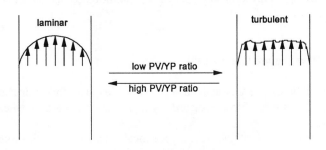

For a Newtonian fluid the onset of turbulence starts if Reynold number Re exceeds a value of 2300.

$$Re = \rho \ x \ v \ x \ (d_o - d_i) / \eta$$

Re	=	Reynold number	[ ]
ρ	=	fluid density	[kg/m³]
v	=	average fluid velocity	[m/s]
d_o	=	open hole diameter	[m]
d_i	=	drill string diameter	[m]
η	=	fluid viscosity	[kg/ms]

2.1 Laminar flow regime (Re < 2300)

Laminar flow of fluids prevails at low flow velocities. All fluid particles move in concentric cylinders parallel to the conduit axis. Adjacent layers of fluid slip past each other with no mixing or interchange of fluid from one layer to the next. Laminar flow can be visualized as a series of concentric cylinders sliding past each other in a manner similar to the tubes of a telescope. The velocity of the cylinders increases from zero at the borehole wall to a maximum at the axis of the borehole, resulting in a parabolic velocity profile (Fig. 1).

Due to this velocity profile v_r is maximum at the centre and minimum at the borehole wall. At the borehole wall, the cuttings are actually moving downward since the fluid velocity is nil there. Due to borehole wall irregularities, rotation of the drill string and impacts by other particles the cuttings are dragged into the centre with high fluid velocity and are eventually lifted in a helical motion towards the surface. [2]

All in all, hole cleaning is not optimum in laminar flow regimes.

Laminar flow is preferred in unconsolidated formations (soft gumbo, unconsolidated sands) because there is only very little impact on the borehole by the moving drilling fluid.

2.2 Turbulent flow regime (Re > 2300)

Turbulent flow of fluids prevails at high flow velocities. The fluid particles move in tumbling or chaotic motion. Flow is disorderly and there is no orderly shear between layers.

A fluid in turbulent flow is subject to random local fluctuations both in velocity and direction while maintaining an average velocity parallel to the axis.

Flow equations relating flow behaviour to the flow characteristics of the fluid are empirical for the turbulent regime. Mathematical description of the laminar flow is based on the Newtonian, the Bingham or Power Law model.

Fig. 1 shows the velocity profile of a Newtonian fluid in turbulent flow. The profile is flatter than the one for the laminar flow profile. Therefore, the carrying capacity is more homogeneous and better than in laminar flow.

Turbulent flow requires more energy which increases circulating pressure when compared to laminar flow. This is due to the chaotic movement of fluid particles as they are lifted out the well. Thus, the danger of washouts, formation damage or loss of circulation is higher for a turbulent flow regime.

2.3 Transition from laminar to turbulent flow regime

The dimensions of the circulation system and the physical properties of the circulating fluid are the two principal parameters affecting the hydraulics of the fluid. For a given wellbore, drilling tools geometry and a well-defined Newtonian drilling fluid, only the increase of pump energy relating to higher flow rate determines the flow regime. As discussed above, at a critical average velocity of the fluid the Reynold number exceeds the value of 2300 resulting in a turbulent flow regime with high carrying capacity for cuttings but also high damaging potential for the borehole wall.

Since real drilling fluids are not Newtonian but behave like Bingham plastics there is a second way to influence the flow regime. Bingham plastics are characterized by two parameters: plastic viscosity (PV) and yield point (YP) (Fig. 2).

The PV of drilling fluids is a measure of the resistance to flow caused by the shearing action of the liquid itself, the mechanical friction between solids, and between the solids and the fluid surrounding them. The higher the PV the higher the pump energy needed to obtain a certain flow rate.

The yield point is an indication of interparticle attraction while the fluid is moving and relates to the carrying capacity of the fluid when in motion. Turbulent flow is favoured at a low PV/YP ratio whereas high PV/YP ratios cause laminar regimes at a given flow rate (Fig. 1).

PV and YP are determined with a Fann® 35 viscometer from the flow curve as follows:

PV = DI (600) - DI (300)   [cP]
YP = DI (300) - PV   [lbs/100 ft²]

DI (600): Fann® 35 dial reading at a shear rate of 600 rpm.

Fig. 2: Bingham model for drilling fluid rheology

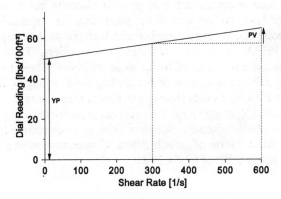

A second approach to describe real drilling fluids is the Power Law model.

$$\tau = k \times \gamma^n$$

τ	=	shear stress	[N/m²]
γ	=	shear rate	[1/s]
n	=	flow behaviour index	[ ]
k	=	consistency index	[Ns/m²]

The   exponent n describes the flow regime, ranges from zero to one and may be calculated as follows:

$$n = 3.32 \times \log_{10} [DI\ (600)/DI\ (300)]$$

A low PV/YP-ratio relates to a low value for n. Such a drilling fluid shows turbulent flow, a "flat" flow profile and good hole cleaning performance at relatively low flow rates.

3. Common Drilling Fluid Viscosifiers

In most cases, bentonite does not provide proper viscosity for hole cleaning therefore additional additives are used.

The high viscosity versions of cellulose polymers, such as polyanionic cellulose (PAC), carboxymethyl cellulose (CMC) and hydroxyethyl cellulose (HEC) are common types of viscosifiers. These cheap materials not only provide viscosity but also fluid loss control. However, they have no or very little shear thinning capabilities. These products only provide good hole cleaning together with high plastic viscosity.

Polyacrylamides (PA, PHPA) are a second class of viscosifiers. These synthetic products have an additional advantage of being shale inhibitive. Their effectiveness declines with increasing salt concentration of the drilling fluid. Along with high YP these materials give high PV which limits their application as viscosifiers.

A third type of viscosifier is Xanthan Gum. This material is shear-thinning providing a high yield point and low plastic viscosity. Xanthan Gum is an expensive biopolymer formed by the bacterial fermentation of carbohydrates. Concentrations of up to 4 ppb (=1.1 %) are needed to obtain a YP of 60 lbs/100 ft.

4. Mixed Metal Hydroxide/bentonite drilling fluids

As a consequence of all the drawbacks of conventional viscosifiers, a new one was looked for which is highly shear-thinning, thermally stable, electrolyte resistant and cost effective.

In the eighties, extensive studies carried out by Burba resulted in a novel drilling fluid viscosifier based on mixed metal hydroxide (MMH) chemistry. [3]

Analytical data of the unique rheological behaviour for a typical system are given by Table 1. The measurement for concentration used by the petroleum industry is pound per barrel (ppb). 1 ppb correlates to 1 g additive in 350 ml solution.

Due to the high carrying capacity (YP), good pumpability (low PV), high gel strengths (GS) and flat flow profile (n = 0.12) MMH compounds are useful inorganic viscosifiers for the drilling industry.

Table 1: Rheology of a typical MMH/bentonite (10 ppb) slurry (pH = 10)

MMH [ppb]	Fann rheology 600-300-200-100-6-3	PV [cP]	YP [lbs/100 ft^2]	GS (10"/10')
0.0	7-5-4-3-2-1	2	3	3/3
0.7	63-58-54-49-30-27	5	53	25/26

1 ppb = 1 g/350 ml

n = 3.32 x log$_{10}$ (63/58) = 0.12

4.1 Mixed Metal Hydroxide

4.1.1 Definition and Physical Properties

Mixed metal hydroxides are white, crystalline, inorganic compounds containing two or more metals surrounded by a hydroxide lattice. MMH crystals are sheet-like, hexagonal platelets with a diameter of 100 nm and a thickness of 2 nm.

4.1.2 Preparation

MMH compounds are preferably prepared by coprecipitation wherein salts of the metals are intimately mixed with alkaline solutions.
A typical MMH compound used in drilling fluids industry is made from magnesium chloride and aluminium chloride:

$$MgCl_2 + AlCl_3 + 4.66 \ NaOH \longrightarrow [MgAl(OH)_{4.66}]^+ \ [Cl]^-_{0.33} + 4.66 \ NaCl$$

4.1.3 Mode of Action

If only magnesium ions were present, all the octahedral gaps of the hydroxide lattice would be occupied by the metal ions and the compound would be electrically neutral (brucite structure). In case only aluminium ions were present, two-thirds of the octahedral gaps of the hydroxide lattice have to be occupied to give an electrically neutral mineral (gibbsite). If both metals are incorporated into the hydroxide lattice irregularities of the lattice occur resulting in an excess of positive charges by the metal ions. This excess of positive charge is neutralized by chloride anions in MMH. All at all, the MMH particles are small, flat crystals with a high positive charge density. When dispersed in a bentonite slurry they form complexes with the much bigger (ca. 2 μm) bentonite particles by coordinating to the negatively charged basal planes and ion exchange mechanism whereby free NaCl is formed. Therefore, MMH can be looked at as a bridging agent for bentonite. Since the mechanism is electrostatic in nature the MMH/bentonite gel forms immediately when shear stress is removed. This also explains why the gel strengths are non-progressive and why the yield point is high.

Since the gel consists of 97 and more per cent of water and behaves like a pseudo-solid it is reasonable to believe that a large amount of the water molecules is structured by the electrostatic force field of the bentonite/MMH complexes. Interaction between MMH and bentonite particles creates that strong electric field which is not observed for the single components. It is also theorized that the MMH/bentonite system forms sheets of associated complexes with coordinated water occupying the spaces between the layers. Application of a certain amount of mechanical energy corresponding to gel strength is needed to break the gel, to turn it from pseudo-solid to water-like liquid. When in motion, only very small energy corresponding to plastic viscosity, is needed to overcome the interlayer attraction or in other words to increase the shear rate between two adjacent layers.

4.2 Rheological characteristics

As indicated by Table 1, a dramatic change in viscosity is observed when only 0.7 g of MMH is added to 350 ml of prehydrated bentonite slurry.

The behaviour of the MMH/bentonite system is quite unusual. At rest, it is like an elastic solid. However, application of little mechanical energy destroys the pseudo-solid and turns it into a water like liquid. When shear stress is removed, the fluid reverses almost instantaneously to the gelled state. Fig. 3 demonstrates that the logarithm of viscosity is in linear correlation to the reciprocal value of the logarithm of shear rate.

Table 2: Shear-thinning of an MMH/bentonite slurry

Shear rate [rpm]	600	300	200	100	6	3
Shear rate [m/s]	1020	510	340	170	10.2	5
Fann® 35 Dial reading	63	58	54	49	30	27
Viscosity multiplier	0.5	1	1.5	3.0	50	100
Viscosity [cP]	31.5	58.5	81	147	1500	2700

Another interesting aspect is that the 10 sec. and 10 min. gel strengths are relatively high and more or less equal. In other words the gel is formed in less than 10 seconds. It is actually assumed that the gel structure is restored in less than 2 μs.

Fig. 3: Shear-thinning of an MMH/bentonite slurry

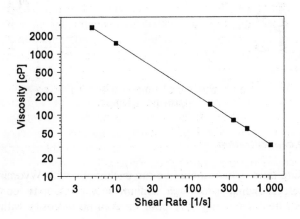

4.3 Variation of parameters [4]

There are many parameters for such a simple system like MMH/clay, e. g., absolute and relative concentrations of bentonite and MMH, pH, temperature, etc.

4.3.1 MMH concentration

The effect of MMH concentration was studied on a bentonite slurry containing 8 ppb (= 2.3 %) Wyoming bentonite. pH was adjusted to 10.0 with soda ash in all cases. Figure 4 shows that plastic viscosity is not much influenced by increasing dosage of MMH but yield point changes dramatically. It is important to exceed a threshold concentration of approximately 0.5 ppb (= 0.14 %) to obtain a stable value for the yield point. On the other hand, overtreatment of the drilling fluid with MMH does not generate additional interaction of system's particles when in motion (= YP) but should be avoided for economical reasons.

Fig. 4: Effect of MMH concentration on fluid rheology

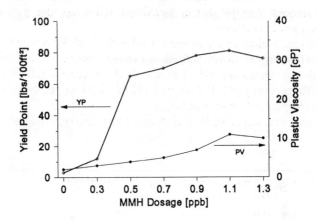

4.3.2 Bentonite concentration

The MMH drilling fluid system should be run exclusively with Wyoming bentonite of high quality corresponding with high sodium montmorillonite content. Peptized (polymer treated) bentonites may give poor or even no viscosity with MMH due to their anionic nature.

Figure 5 shows in summary that the higher the bentonite content the higher the obtainable viscosity (YP) for a freshwater mud.

Fig. 5: Effect of bentonite concentration on fluid rheology

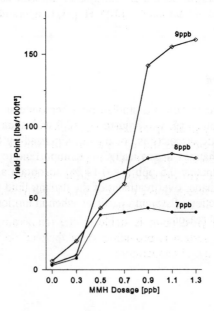

4.3.3 pH

It was experienced that pH has a significant effect on the performance of the MMH/bentonite drilling system.
The influence was studied on a freshwater mud with 7 ppb Wyoming bentonite and 0.7 ppb MMH. The pH was varied with soda ash and caustic soda. The resulting yield points are plotted versus the pH in Figure 6. As a general rule, the yield point is increasing with increasing pH. But above a pH of 10 the increase in yield point is only minor; no further viscosity gain is observed.
The most economical use of MMH therefore is to run the system at pH 10.5.

Fig. 6: Influence of pH on MMH/bentonite fluid rheology

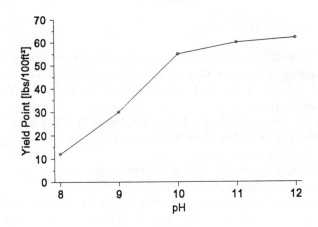

4.3.4 Temperature

MMH by itself is thermally stable up to 250°C and higher.
The stability of the yield point was studied on a bentonite/MMH system consisting of 8 ppb Wyoming bentonite and 0.9 ppb of MMH. The pH was adjusted to 11 with soda ash. The mud was hot-rolled for 16 hours at different temperatures and cooled to room temperature. After readjustment of the pH to 11 with soda ash, rheology was determined with a FANN® 35 SA viscometer.
Figure 7 shows that the yield point is stable in the temperature range from 25 to 205°C. That means the system is providing shear-thinning and good hole cleaning properties over a wide range of temperature.

Fig. 7: Influence of temperature on MMH/bentonite fluid rheology

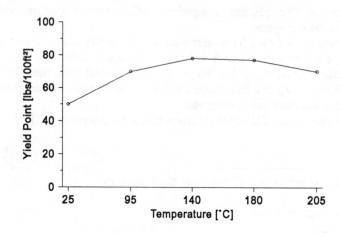

4.3.5 Electrolyte tolerance

Once the bentonite/MMH gel has been formed, any common type of salt can be added without destroying the shear-thinning properties of the fluid. It is, however, impossible to achieve viscosity when MMH is added to a salted system. As a consequence, it is mandatory to mix MMH into a freshwater system.

Seawater, NaCl and KCl up to saturation and calcium/magnesium up to 100,000 ppm are tolerated by bentonite/MMH systems.

In case of electrolyte contamination approximately double the bentonite concentration is required to achieve the same viscosity as in a freshwater system. As a rule, the bentonite concentration required increases with the salt content.

4.3.6 Contaminants

MMH interacts with bentonite particles by means of its cationic charge. This mechanism may be disturbed by anionic polymers.

Additions of as little as 0.1 - 0.2 ppb lignite, lignosulfonate, CMC or PAC may result in a collapse of YP and gel strenghts.

The unpredictable behaviour of strongly anionic products in the MMH system prohibits their use for fluid loss control. Only non-ionic polymers may be used for fluid loss control.

Table 3: Thinning of an MMH mud by anionic additives

Additive	Dosage [ppb]	PV [cP]	YP [lbs/100 ft^2]	GS (10"/10')
blank	-----	5	53	25/26
Na-polyacrylate	0.2	4	2	2/2
PAC-LV	1	6	6	3/3

5. Applications in drilling

The unique rheological performance (n < 0.2, PV < 10, YP > 50; shear thinning, non-progressive gel strengths) of MMH/bentonite slurries has many advantages in drilling fluid technology.

5.1 Milling

Milling is the removal of a set steel casing from the well by special tools which cut the casing into small chunks. It has to be done to side track from an existing well or when the set casing is corroded.

The excellent carrying capacity of an MMH fluid (high YP) makes it ideal for milling wells. Effective removal of metal cuttings has been experienced in more than one hundred jobs. Another major advantage of this milling fluid is that it requires less pump pressure for efficent flow compared to conventional muds with similar viscosities. Thus, less energy is needed and the milling rate is higher. Record milling jobs with 50 % reduced milling time have been reported.

A milling fluid consisting of 10 ppb bentonite and 1 ppb MMH suspends metal pieces and lumps of concrete weighing several kilograms when at rest. Settling of the cuttings is not observed even after a long period of time. After a period of rest, e. g., round trip, only low pump pressure is required to break the gel (non-progressive gel strength). All in all, MMH/bentonite fluids are the most advanced milling fluid system available to the industry today.

5.2 Directional and horizontal drilling [5,6,7,8]

Deviated and horizontal drilling have been developed to a state of maturity in the 80 's and 90 's. The essential reason for this development is that horizontal wells produce on average four times more than vertical wells because of the greater production area exposed to the wellbore.

Due to the wellbore geometry it was and still is a challenge for the drilling fluid industry to provide suitable drilling fluids.

Since the distance a cutting can fall before reaching the wellbore wall is decreasing as the angle of the hole increases hole cleaning is always a concern in high angle and especially in horizontal holes. [5] Problems caused by poor hole cleaning are stuck pipe, high torque and drag, low rate of penetration etc.

Laboratory studies under simulated well conditions by Okrajni and Azar show that a high YP/PV-ratio is advantageous for hole cleaning especially at low annular velocities.[6,7]

Since MMH/bentonite muds do have high low shear rate viscosities, they flow as a solid mass carrying the cuttings out of the hole. A successful application of the MMH/bentonite system in horizontal drilling offshore Lousiana is reported by Polnaszek and Fraser. [7,8] The drilling was not only horizontal but also in unconsolidated sand of high permeability. The drilling fluid base was water/seawater, 10 ppb bentonite and 1 ppb MMH and was adjusted to pH 10 with caustic soda to provide suitable rheology. The fluid was weighted up to 1.26 S. G. with grounded marble which also provided spurt loss control for the permeable sand. The annular fluid velocity was about 20 m/min. and the horizontal section of 290 meters was drilled at an average rate of penetration of 50 m/h. The authors stress that the hole was always clean, open and gauge. Due to the minimal impact of the grounded marble treated MMH drilling fluid production is high without using any stimulation technique.

5.3 Drilling in unconsolidated formations [8,9]

When pumping MMH fluid through a well, a highly viscous film of fluid can be observed near the borehole wall. This is verified by the fact that cuttings can reach surface in only three fourths of the calculated bottoms up time based on gauge hole.[8]

The fluid actually moving must flow faster than the calculated average flow rate explaining the shorter than theory bottom up time. Due to the shear-thinning properties of the MMH system the fluid at the borehole wall is almost static. This separates the borehole wall from the moving fluid which has a stabilizing effect on unconsolidated formations.[9]

6. Conclusions

MMH fluids show novel and unique rheological properties. Nonprogressive gel strengths, high yield points, shear-thinning and low plastic viscosity are explained by an electrostatic mechanism.

Benefits achievable with MMH drilling fluids include:

- optimum cuttings transport and clean borehole
- good solids suspension
- high rate of penetration
- stabilization of unconsolidated formations
- high return permeabilities and
- lower drilling costs

MMH fluids have been used successfully in field applications, as follows:

- milling jobs
- horizontal drilling
- drilling through unconsolidated sand formations
- well completion and
- civil engineering, tunneling

Last but not least, because of its inert inorganic nature and poor solubility in water, MMH compounds are nontoxic and environmentally benign.

7. Acknowledgements

The author would like to thank SKW Trostberg AG for permission to publish this article. Special thanks go to Johann Plank who gave valuable contributions in numerous discussions.

8. References

1. NL Baroid/NL Industries, Inc., Manual of Drilling Fluids Technology, "Calculations, Charts, and Tables for Mud Engineering", 1985, p. 58-70.

2. H.C.H. Darley, G. R. Gray, "Composition and Properties of Drilling and Completion Fluids", 5th edition, 1991, p. 258-267.

3. J.L. Burba, "Laboratory and Field Evalutations of Novel Inorganic Drilling Fluid Additive", <u>IADC/SPE 17198</u>, presented in Dallas, Texas, March 1988.

4. <u>SKW Technical Brochure</u>, "POLYVIS® - Inorganic Drilling Fluid Viscosifier", January 1993.

5. R. Caenn, G. West, "A Comparison of Water Base vs. Oil Base Mud for Horizontal Drilling and Extended Reach Drilling", <u>Drilling Fluids Technology Conference</u>, presented in Houston, Texas, April 1992.

6. S. S. Okrajni, J. J. Azar, "Mud Cuttings Transport in Directional Well Drilling", <u>SPE 14178</u>, Las Vegas, USA, September 1985; and SPE Drill. Eng., August 1986, p. 291.

7. L. J. Fraser, "Drilling Fluids Considerations and Design for Horizontal Wells", 4th International Conference on Horizontal Well Technology, presented in Houston, Texas, October 1992.

8. L. C. Polnaszek, L. J. Fraser, "Drilling Fluid Formulation for Shallow Offshore Horizontal Well Applications", <u>SPE 22577</u>, presented in Dallas, Texas, October 1991.

9. F. Lavoix, M. Lewis, "Mixed Metal Hydroxide Drilling Fluid Minimizes Well Bore Washouts", <u>Oil & Gas Journal</u>, September 28, 1992, p. 87.

The Use of Fourier Transform Infrared Spectroscopy to Characterise Cement Powders, Cement Hydration and the Role of Additives

T. L. Hughes, C. M. Methven, T. G. J. Jones, S. E. Pelham, and P. Franklin

SCHLUMBERGER CAMBRIDGE RESEARCH, HIGH CROSS, MADINGLEY ROAD, CAMBRIDGE CB3 0HG, UK

1. INTRODUCTION

The analysis of the mineral phases in cement clinkers, cement powders and setting cement samples has proven to be a difficult and time-consuming task. The commonest methods for quantifying the so-called Bogue phases in cement clinkers are light microscopy[1], X-ray diffraction[2] and chemical analysis of the oxide content[3]. These methods have been compared[4]. A complete analysis of the phase composition of cement powders should also include the calcium sulphate minerals (gypsum with bassanite and/or anhydrite) and the products formed by the ageing of the cement (calcium hydroxide, calcium carbonate, syngenite, etc.).

Similar difficulties exist in the monitoring of the hydration reactions in a setting cement slurry. The routine monitoring of cement hydration by most chemically specific techniques generally requires samples of setting cement to be quenched (hydration reactions stopped by extracting water) and the residual solid dried and ground. There are few *in situ* techniques which enable the hydration reactions in cement slurries to be monitored with any degree of chemical specificity. Commonly-used *in situ* techniques such as heat-flow calorimetry[5] and rheometric thickening profiles[6] are sensitive to hydration reactions but have little chemical specificity. Recently, Barnes and coworkers[7] have used energy-dispersive synchrotron radiation as a source of X-rays for transmission X-ray diffraction which has allowed cement hydration reactions to be monitored at time intervals of about 10 seconds.

The objective of this paper is to demonstrate the application of Fourier transform infrared (FTIR) spectroscopy to the quantitative analysis of mineral phases in cement powders and to the *in situ* monitoring of the hydration reactions in a cement slurry from mixing to set. The quantitative analysis of cement powders is achieved by the use of FTIR diffuse reflectance spectroscopy[8,9], a technique which is able to detect the bulk Bogue phases, the various calcium sulphate minerals and products of cement ageing. The diffuse reflectance FTIR spectra are interpreted quantitatively using a multivariate calibration model. The *in situ* monitoring of cement hydration is achieved using FTIR attenuated total reflectance (FTIR/ATR) spectroscopy[10], a technique which is sensitive to the surface of the hydrating grains. The time evolution of the FTIR/ATR spectra show the conversion of tricalcium silicate to C-S-H gel, calcium sulphate to ettringite, the formation of calcium hydroxide and the consumption of free water.

2. DIFFUSE REFLECTANCE SPECTRA OF CEMENT POWDERS

(a) Experimental

Diffuse reflectance spectra were collected using a Nicolet 5DX FTIR spectrometer equipped with a Spectra-Tech Collector™ diffuse reflectance accessory. Fig. 1 shows a schematic of the accessory which has been modified to allow the sample cup to rotate during spectral collection, enabling a more representative spectrum to be collected.

Spectra of cement powders were collected from samples diluted to 10 wt percent with KBr ground to a fixed particle size distribution; the cement powders themselves were not ground prior to spectral analysis. The sample cups were filled by compacting 0.4500 g of the cement-KBr mixture using a small compaction cell. Spectra were collected using a TGS detector from 176 scans (4 minute collection time); the spectra were ratioed against a background of the compacted KBr. The resolution of the spectra was 4 cm^{-1}.

(b) Diffuse Reflectance Spectra of Cement Powders

Fig. 2 shows two repeat diffuse reflectance spectra of a cement powder. The spectra were obtained on two completely separate preparations of the same primary cement sample. The difference between the two replicate spectra is within 0.01 absorbance units across the entire spectrum.

The diffuse reflectance spectra show the presence of the mineral phases alite (935, 520 cm^{-1}), belite (990, 878, 847, 505 cm^{-1}), gypsum (3550, 3400, 1683, 1620, 1140, 667, 600 cm^{-1}), bassanite (3611, 3555 cm^{-1}), calcium hydroxide (3644 cm^{-1}), syngenite (3310, 1195 cm^{-1}) and calcium carbonate (broad band in region 1580-1330 cm^{-1}; the frequent occurrence of a split carbonate band with peaks at 1485 and 1415 cm^{-1} indicates the presence of vaterite[11]). The characteristic absorption bands due to ferrite and aluminate are located in the spectral region 800-580 cm^{-1} where they are convolved with the sulphate and silicate absorption bands.

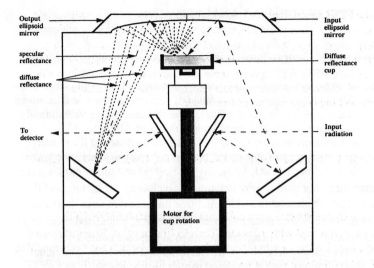

FIGURE 1. Schematic of diffuse reflectance cell.

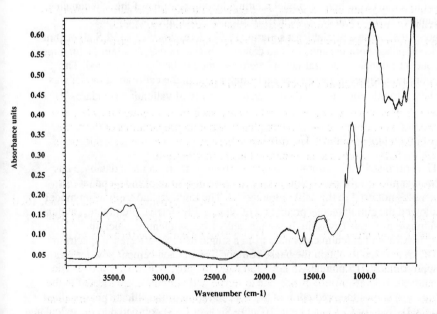

FIGURE 2. Repeat diffuse reflectance spectra of a class G cement.

3. QUANTITATIVE CEMENT ANALYSIS

(a) Composition of Cement Powders and Construction of Calibration Model

A cement database consisting of the composition and diffuse reflectance FTIR spectra of 156 samples, chosen from a range of cement suppliers and locations throughout the world, has been established. The cements were predominantly API class G cements, although a significant number of API class A (construction) cements were included to extend the variance of the model. The bulk oxide composition of each calibration cement was determined using ICP analysis. Both major (CaO, SiO_2, Al_2O_3, Fe_2O_3) and minor oxides were determined; 10 minor oxides were determined, including SO_3, MgO and P_2O_5. The loss on ignition, insoluble residue and free lime content were determined by ASTM methods[12]. The nominal clinker phases were calculated according to API specification[13]. It was assumed that the total sulphate content was partitioned between gypsum, bassanite, anhydrite and syngenite. Individual sulphate concentrations were determined using ratios of isolated characteristic bands in the diffuse reflectance spectra (e.g., O-H band at 3611 cm^{-1} for bassanite) subject to the constraint of the bulk SO_3 and K_2O content. Calcium carbonate and calcium hydroxide concentrations were determined by thermal gravimetric analysis (TGA).

A quantitative calibration model relating the composition of the 156 cement samples in the database to their FTIR diffuse reflectance spectra was generated using a partial least squares (PLS) regression[14]. The PLS algorithm decomposed both the spectral and concentration data matrices into factors; the concentration and spectral factors were assumed to be linearly related. The optimum number of factors for each component in the calibration model was chosen to minimise the prediction error of a suite of validation standards[14].

(b) Prediction of Cement Composition

The optimised PLS calibration model derived from the cement database has been applied to the prediction of composition of a suite of independent test samples not used in the calibration model. The multivariate calibration model allowed the simultaneous prediction of alite, belite, ferrite, aluminate, calcium sulphate minerals (gypsum, bassanite, anhydrite, syngenite), calcium hydroxide and calcium carbonate from a single diffuse reflectance spectrum. The time taken to obtain the phase composition of each cement powder was approximately 15 minutes.

Fig. 3(a) shows the prediction of the concentration of the silicate phases alite (($CaO)_3SiO_2$ or C_3S) and belite (($CaO)_2SiO_2$ or C_2S) compared to the calculated API Bogue concentrations. The bulk of the predicted concentrations are within ± 5 wt percent. An interesting feature of the predicted concentrations is the marked negative correlation between the alite and belite concentrations; this correlation can clearly be seen in fig. 3(a) where

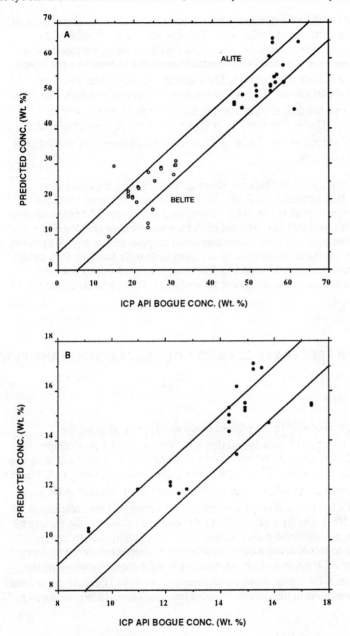

FIGURE 3. Prediction of (A) alite and belite and (B) ferrite in test cements. Lines show position of +/- 5% and 1% error bands, respectively.

overestimates of the alite concentration are accompanied by underestimates of the belite concentration and *vice versa*. This correlation is a feature of the alite and belite concentrations in the cement database; the correlation is also evident between the alite and belite concentrations determined by both light microscopy and X-ray diffraction[4]. The calibration model gives excellent predictions of total silicate ($C_3S + C_2S$) content of cement powders. Fig. 3(b) shows the corresponding comparison for the ferrite phase (tetracalcium aluminoferrite, $(CaO)_4Al_2O_3Fe_2O_3$ or C_4AF). The ferrite content of many of the test samples is predicted to within ±1 wt percent; in all cases it is predicted to within ± 2 wt percent.

The minor (non-clinker) phases are also well predicted by the calibration model. Fig. 4(a) compares the prediction of the concentration of bassanite obtained by single peak height with values predicted by the calibration model. The bassanite concentrations of most of the test samples are predicted to within ± 0.5 wt percent of the values estimated by peak height. Fig. 4(b) shows the prediction of the concentration of calcium hydroxide from the PLS model with the concentration determined by TGA; the predicted values from the diffuse reflectance spectra are almost all within ± 0.25 wt percent of the values from TGA.

4. FTIR/ATR SPECTRA OF SLURRIES OF NEAT CEMENTS AND PURE PHASES

(a) Experimental

All cement slurries used in the present study were prepared using the standard API method[13], namely, mixing in a Waring blender at 4000 rpm for 15 seconds followed by 20000 rpm for 35 seconds. The water/cement ratio was 0.5 for all cement slurries. FTIR/ATR spectra were collected using a Nicolet 800 FTIR spectrometer and using either a 70° zinc selenide horizontal ATR plate (fig. 5(a)) or a light pipe ATR probe with a 2-reflection 45° zinc selenide ATR crystal (fig. 5(b)). The light pipe ATR probe was used to collect the spectra of cement slurries immediately after mixing but before setting; the light pipe spectra were collected using a high-sensitivity MCT detector cooled by liquid nitrogen. The ATR spectra of cement slurries to set were collected using the 70° horizontal ATR crystal using an ambient temperature TGS detector; a small metal liner was used in the sample trough to remove cement after setting (fig. 5(a)).

b) Neat Cements

Fig. 6(a) shows the time evolution of the FTIR/ATR spectra of a neat API class G cement from shortly after mixing to set at ambient temperature. The spectral region of interest is 4000-800 cm^{-1}; below 800 cm^{-1} the absorbance of both water and zinc selenide become very large. For times up to about 10 hours, the

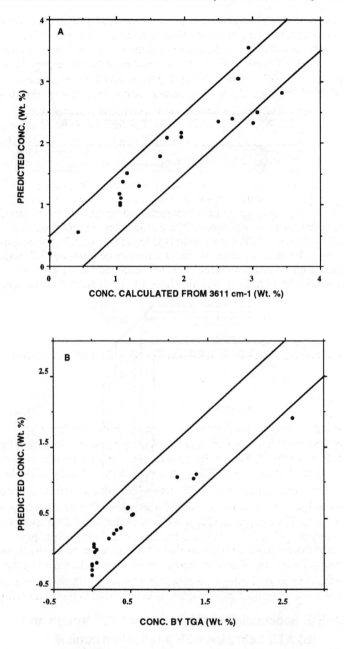

FIGURE 4. Prediction of (A) bassanite and (B) calcium hydroxide in test cements. Lines show position of +/- 0.5% and 0.25% error bands, respectively.

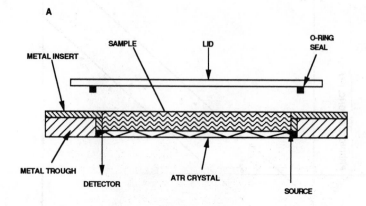

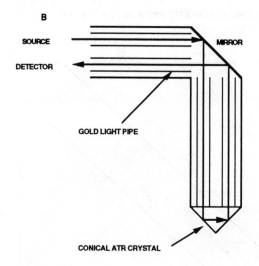

FIGURE 5. Schematics of (A) horizontal ATR trough and (B) ATR light pipe with 2-reflection conical ATR crystal.

A

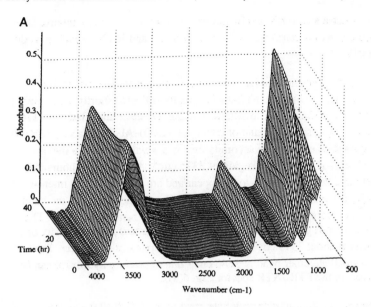

B

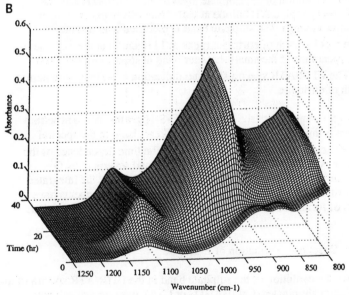

FIGURE 6. Evolution of FTIR/ATR spectra during the hydration of a class G cement slurry: (A) spectral region 4000-800 cm-1; (B) spectral region 1250-800 cm-1.

FTIR/ATR spectra are dominated by the spectrum of water (broad, intense O-H stretching band over spectral region 3700-2700 cm^{-1} and H-O-H bending mode at 1635 cm^{-1}).

The spectra of the cement mineral phases are evident in the spectral region 1250-800 cm^{-1} (fig. 6(b)). The FTIR/ATR spectra collected shortly after mixing show the absorption bands due to alite at 918 and 891 cm^{-1} and a band at 1110 cm^{-1} which is the v_3 sulphate band of ettringite ((CaO)$_3$Al$_2$O$_3$.3CaSO$_4$.32H$_2$O). The early growth of ettringite is accompanied by a small increase in the intensity of the O-H band in the region of 3375 cm^{-1}. After about 3 hours the bands at 918 and 891 cm^{-1} begin to disappear and are replaced by an intense band at 945 cm^{-1} which is characteristic of C-S-H gel. The growth of the band at 945 cm^{-1} is accompanied by the growth of a smaller band at 815 cm^{-1}. The production of C-S-H gel is also accompanied by a decrease in the intensity of the O-H band at about 3375 cm^{-1} due to the consumption of free water. The growth of a small shoulder band at 3640 cm^{-1} due to calcium hydroxide can be discriminated in the FTIR/ATR spectra (fig. 6(a)).

Fig. 7 shows the evolution of the sulphate band in the FTIR/ATR spectra obtained with the light pipe ATR probe at early time (t<8 hours starting at about 1 min after mixing). The sulphate band is modified over a short time indicating the rapid growth of ettringite (peak at 1110 cm^{-1}) at the expense of gypsum (disappearance of the small shoulder band in spectral region 1160-1140 cm^{-1}). The FTIR/ATR spectra show no indication of an induction period for the growth of ettringite.

A convenient representation of the kinetics of C-S-H gel formation in setting cement slurries is the change in the area of the silicate band in the spectral region 1050-900 cm^{-1} with time. Fig. 8 compares the time dependence of the band area for an API class A (construction) cement and several class G cements. The curves, which have the same sigmoidal shape as the integrated heat flow calorimetry curves, show that, although the class G cements have different rates of formation of C-S-H gel, their induction periods are similar.

(c) Pure Phase Reactions: Formation of Ettringite

Fig. 9(a) shows the evolution of the sulphate band in the FTIR/ATR spectra of a paste of tricalcium aluminate ((CaO)$_3$Al$_2$O$_3$ or C$_3$A), gypsum and calcium hydroxide. The ratio of these three phases (by weight) was 2:1:1 as used by Gaidis and Gartner[15]; the water:solid ratio was unity.

The first 26 hours of the reaction was characterised by the growth of ettringite (peak at 1110 cm^{-1}) and the disappearance of gypsum (shoulder band in the spectral region 1160-1140 cm^{-1}). The weak v_1 sulphate band of ettringite was observed at 989 cm^{-1} after about 21 hours; the second v_1 band at 1005 cm^{-1} was not observed. After 26 hours the band at 1110 cm^{-1} was

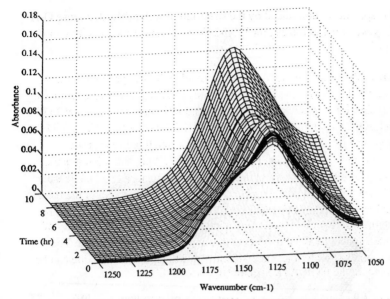

FIGURE 7. Early-time evolution of FTIR/ATR spectra of class G cement; spectra collected with light pipe ATR probe

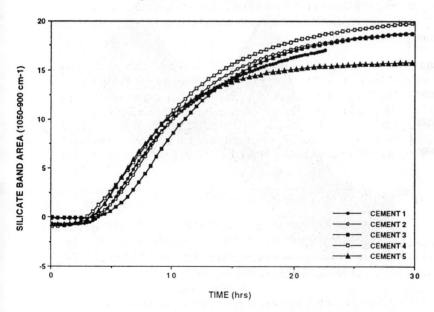

FIGURE 8. Comparison of time dependence of silicate band area (1050-900 cm-1) from FTIR/ATR spectra of 4 class G cements (1-4) and a class A cement (5).

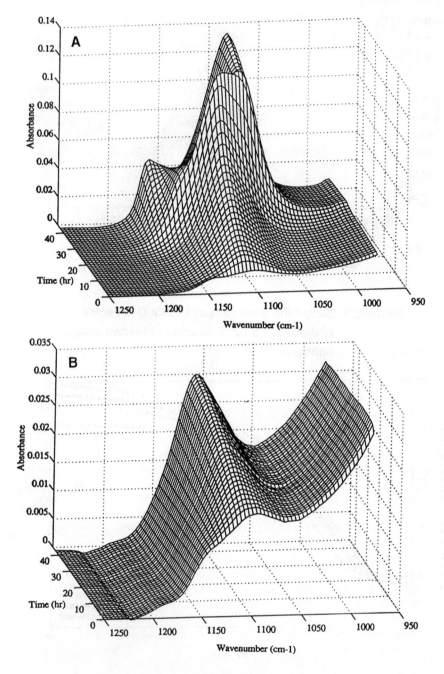

FIGURE 9. Evolution of sulphate band in FTIR/ATR
spectra of (A) C3A/CaSO4.2H2O/Ca(OH)2;
(B) C4AF/CaSO4.2H2O/Ca(OH)2.

increasingly replaced by bands at 1160 and 1083 cm^{-1} which are the characteristic of the v_3 sulphate band of the monosulphate phase $((CaO)_3Al_2O_3.CaSO_4.11H_2O)$. The appearance of the monosulphate phase was accompanied by one of the weak v_1 sulphate bands at 982 cm^{-1}; the other weak v_1 sulphate band at 992 cm^{-1} did not appear in the FTIR/ATR spectra. Kuzel and Pöllmann[16] have followed the formation of ettringite and monosulphate using an *in situ* X-ray diffraction technique. The ratio of the three phases used by Kuzel and Pöllmann was approximately 3.6:2.3:1 by weight, giving a higher sulphate:aluminate ratio than in the present study. The X-ray diffraction technique showed the first appearance of monosulphate after 65 hours; the later appearance of monosulphate than in the present study is presumably due to the higher sulphate:aluminate ratio.

Fig. 9(b) shows the corresponding evolution of the v_3 sulphate band in a paste of ferrite, gypsum and calcium hydroxide (weight ratio 2:1:1 with water:solid ratio of unity). The rate of growth of ettringite was considerably slower in the paste containing ferrite and after 40 hours monosulphate had not been detected.

5. THE EFFECT OF SOME CARBOHYDRATE ADDITIVES ON THE GROWTH OF ETTRINGITE

It is well known that many carbohydrates such as sugars, aliphatic polyols and aromatic polyphenols are powerful retarders of silicate hydration[7,17]. Their use as retarders in oilwell cement slurries has been limited by problems such as sensitivity to additive concentration and a tendency for early-time gelation of the slurries making them difficult to pump. The problem of gelation appears to be related to the modification of the hydration of the interstitial phase by the presence of the carbohydrate additives. If the carbohydrate retarders are interacting with the interstitial phases, then it is expected that the retarders will influence the production of ettringite.

Fig. 10 compares the relative rates of growth of ettringite in a neat class G cement slurry and in the presence of a number of carbohydrate retarders. The curves show the absorbance at 1110 cm^{-1} measured relative to a local baseline set at 1250 cm^{-1}; FTIR/ATR spectra were collected with the light pipe ATR probe. Within the first few tens of minutes all of the carbohydrate additives retarded the formation of ettringite relative to that formed in the neat cement. However, after about 1 hour many of the carbohydrate additives had promoted the growth of ettringite relative to the neat slurry. The slurries in which the ettringite was promoted by the carbohydrate gelled and produced no free water. In contrast, the sugars raffinose and stachyose and the aromatic diphenol catechol at a concentration of 0.5 weight percent retarded the formation of ettringite relative to the neat slurry for times in excess of 2 hours. The slurries containing raffinose and stachyose were of markedly lower viscosity than the neat slurry and the production of free water by these retarded slurries was similarly greater. However, the slurry with 0.5 weight percent catechol gelled severely and produced no free water despite the reduced ettringite formation.

The results suggest that the carbohydrates, with the exception of stachyose and raffinose, promoted the dissolution of the interstitial phases. The increased dissolution of the interstitial phases resulted in the increased formation of ettringite and the formation of calcium aluminate hydrates (most probably $(CaO)_3Al_2O_3.6H_2O$ or C_3AH_6), the latter causing the gelation of the slurry. The calcium sulphate in the cement was therefore inadequate to control the hydration of the interstitial phases promoted by the carbohydrate additives. When the concentration of catechol reached 0.5 weight percent, the precipitation of ettringite appeared to be retarded but dissolution of the interstitial phases was still promoted since the slurry was severely gelled. The sugars raffinose and stachyose were exceptional in that they did not promote the dissolution of the interstitial phases but appeared either to retard the dissolution of the interstitial phases or to retard the formation of ettringite.

It has been observed[18] that the sugar raffinose is a powerful retarders of silicate hydration; stachyose is also a powerful retarder. Raffinose and stachyose are more powerful retarders than other sugars such as sucrose or glucose. One possible explanation for the retarding power of raffinose and stachyose is that their inability to promote the dissolution of the interstitial phases results in more of the additive being available to retard the hydration of the silicate phases. The promotion of the dissolution of the interstitial phases by carbohydrates such as sucrose and mannitol probably results in their consumption and/or modification which renders them less effective as silicate retarders.

6. DISCUSSION

Lacking knowledge of the true compositions of the clinker phases in the calibration model, the quantitative method makes explicit use of a Bogue transform between bulk oxide composition and the nominal API Bogue clinker phase composition. The calibration model is therefore predicting the concentration of invariant Bogue phases from variant spectral features in the spectral data. Samples identified by the calibration model to be outliers may have either a composition which is genuinely outside the composition range of the model or an anomalous clinker phase composition. The problem of identifying the composition (or range of compositions) of the clinker phases is not restricted to infrared spectroscopy. Quantitative techniques such as X-ray diffraction also require an assumption of the composition (and measurement response) of the clinker phases. The use of Rietveld methods with quantitative X-ray diffraction[2] offers a method of identifying the composition of the average clinker phases. The use of FTIR microscopy[19] may be a promising method for obtaining the spectral variance of the clinker phases down to a length scale of about 10 microns.

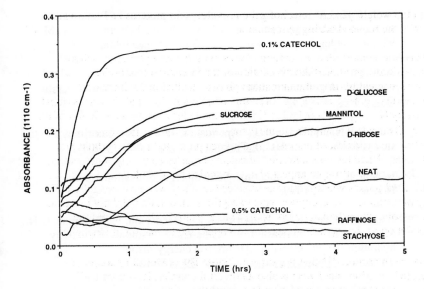

FIGURE 10. Time dependence of absorbance at 1110 cm-1 from
FTIR/ATR spectra of class G cement slurries
containing various carbohydrate additives.

FTIR/ATR spectroscopy for *in situ*, chemically-specific chemical monitoring of cement hydration appears to be a promising technique. The ATR technique yields information on the formation of C-S-H gel, early-time sulphate composition and consumption of free water and may be a good complement to *in situ* X-ray diffraction techniques[7,8]. The early-time detection of the formation of C-S-H gel is comparable to heat flow calorimetry and the time scales over which the two techniques detect the rate of cement hydration are comparable. The ATR technique is less useful in determining the formation of calcium hydroxide and the hydration products of the interstitial phases other than ettringite and monosulphate. Extension of the *in situ* FTIR/ATR technique to elevated temperatures and pressures is currently being pursued.

7. REFERENCES

1.   Campbell, D.H., 'Microscopical examination and interpretation of Portland cement and clinkers', Portland Cement Association, Skokie, Illinois, 1986.

2.   Taylor, J.C. and Aldridge, L.P., "Full-profile Rietveld quantitative XRD analysis of Portland cement: XRD profiles for the major phase tricalcium silicate (C_3S:$3CaO.SiO_2$)", <u>Powder Diffraction</u>, 1993, <u>8</u> 138-144.

3.   Scott, E.H., "Atomic absorption methods for analysis of Portland cement" in 'Rapid methods for chemical analysis of hydraulic cement', ASTM STP 985, R. Gebhardt (editor), pp 15-25, ASTM. Philadelphia, 1988.

4.   Aldridge, L.P., "Accuracy and precision of phase analysis in Portland cement by Bogue, microscopic and X-ray diffraction methods", Cem. Concr. Res., 1982, 12 381-398.

5.   Aukett, P.N. and Bensted, J., "Application of heat flow calorimetry to the study of oilwell cements", J. Thermal Analysis, 1992, 38 701-7.

6.   Nelson, E.B., 'Well Cementing', Schlumberger Educational Services, Houston, Texas, 1990.

7.   Barnes, P., Clark, S.M., Hausermann, D., Henderson, E., Fentiman, C.H., Muhamad, M.N. and Rashid, S., "Time-resolved studies of the early hydration of cements using synchrotron energy-dispersive diffraction", Phase Transitions. 1992, 39 117-128.

8.   Griffiths, P.R. and Fuller, M.P., "Mid-infrared spectrometry of powdered samples" in 'Advances in infrared and Raman spectroscopy', Vol. R.J.H. Clark and R.E. Hester (editors), Heyden, London, 1982.

9.   Chalmers, J.M. and Mackenzie, M.W., "Solid sampling techniques" in 'Advances in applied Fourier transform infrared spectroscopy', M.W. Mackenzie (editor), John Wiley, Chichester, 1988.

10.   Harrick, N.J., 'Internal reflection spectroscopy', Wiley-Interscience, New York, 1967.

11.   White, W.B., "The carbonate minerals" in 'The infrared spectra of minerals', V.C. Farmer (editor), The Mineralogical Society, London, 1974.

12.   ASTM, "Standard chemical methods for chemical analysis of hydraulic cement" in '1990 Annual book of ASTM standards', volume 04.01, C-144, ASTM, Philadelphia, 1990.

13.   API, 'API specification for materials and testing for well', API Spec. 10, 2nd edition, Dallas, 1984.

14.   Martens, H. and Naes, T., 'Multivariate calibration', Wiley, Chichester, 1989.

15.   Gaidis, J.M. and Gartner, E.M., "Hydration mechanisms II" in 'Materials science of concrete II', J. Skalny and S. Mindess, editors, American Ceramic Society, Westerville, Ohio, 1991.

16.   Kuzel, H-J and Pöllmann, H., "Hydration of C_3A in the presence of $Ca(OH)_2$, $CaSO_4.2H_2O$ and $CaCO_3$", Cem. Concr. Res., 1991, 21 885-895.

17.   Taleb, H., 'Analytical and mechanistic aspects of the action of selected retarders on the hydration of "tricalcium silicate", the major component of Portland cement', PhD thesis, University of Georgetown, Washington DC, 1985.

18.   Wilding, C.R., Walter, A. and Double, D.D., "A classification of inorganic and organic admixtures by conduction calorimetry", <u>Cem. Concr. Res.</u>, 1984, <u>14</u> 185-194.

19.   Caveney, R. and Price, B., "FT-IR microscopy spectroscopy study of cement crystal phases", Proc. 14th Int. Conf. Cement Microscopy, Costa Mesa, California, April 1992, pages 114-133.

The Controls on Barium Concentrations in North Sea Formation Waters

E. A. Warren and P. C. Smalley

BP EXPLORATION, CHERTSEY ROAD, SUNBURY-ON-THAMES, MIDDLESEX, TW16 7LN, UK

ABSTRACT

Barium-rich formation waters cause the formation of barium sulphate scales in North Sea oil production. These scales are very costly to remove. A new compilation of North Sea formation water data has enabled barium concentrations to be mapped spatially across the North Sea. This reveals two main barium "hotspots": in Quad 16 fields in the central North Sea, and in Norway's Haltenbanken. The barium-rich waters are statistically discrete chemical compositions to other North Sea formation waters. They are potassium rich and magnesium poor and have high bicarbonate.

All barium rich waters in Quad 16 are found in reservoirs from a single geological formation: the late Jurassic Brae formation. Although barium concentration does not appear to have been acquired within the reservoir, it has been acquired locally. Shales are a possibility.

Knowledge of barium concentration and reservoir geology may thus enable predictions of likely barium concentration to be made locally.

INTRODUCTION

Barium sulphate scale is a major problem in North Sea oil production operations because it precipitates in pipework and reservoir rock perforations so reducing flow. It forms by the mixing of barium-rich connate formation water from the reservoir with sulphate-rich sea water introduced into the well in drilling and completion fluids, or by breakthrough of injection waters used for pressure support. Once formed, it is difficult and costly to remove. Prediction of the susceptibility of any well to barium sulphate scale formation is clearly desirable in order to optimise engineering of the facilities required and the reservoir production plan so minimising costs. This requires knowledge of the barium concentration of the formation water in the reservoir. Unfortunately, this requires a representative sample of the formation water for the field to be obtained which, for many fields, due to operation difficulties or excessive well costs, is often not possible. Another possibility of determining barium concentration is by prediction. However, this requires detailed knowledge of the spatial variations in barium concentration in formation water compositions and their geochemical controls. This paper presents the results of a new compilation of formation water compositions which has enabled a map of to be constructed for the first time of barium concentrations across the North Sea basin. The controls on the distribution of barium-rich waters are then investigated in an attempt to produce a predictive model for barium concentration.

APPROACH

220 chemical analyses of formation waters considered representative of the reservoir were provided by 15 operators of oil and gas fields throughout the North Sea (Warren & Smalley in press). The data were analysed with multi-variate statistical tools, correlations and standard cross-plots. The data were then compiled into a map of barium concentration for the North Sea basin.

No attempt was made to contour the data because there is no direct evidence of concentration gradients between fields, and much evidence of compartmentalisation resulting in large changes in fluid compositions across faults and barriers. Variations in barium concentrations are illustrated based on the frequency distribution of the data. Class subdivisions were assigned according to the 10th, 25th, 50th, 75th and 90th percentiles of the data. This approach is widely used in geochemical prospecting because it produces a balanced, statistical illustration of the variability within a dataset. Although a linear scale is simpler to construct, both the number of subdivisions and the class intervals are chosen subjectively. Furthermore, a linear subdivision tends to exaggerate outliers and obscure variability in the majority of the data.

RESULTS

Barium concentrations vary by over three orders of magnitude in produced formation waters in the North Sea, from less than 5 mg/l to over 2500 mg/l. Examples are given in Table 1.

The map of barium concentration (figure 1) illustrates spatial variations across the North Sea basin. Two major barium hotspots occur, in Quad 16 fields in the central North Sea, and in Haltenbanken. Barium concentrations are lowest in the sulphate-rich waters of the southern North Sea gas fields. It is clear, however, that there are very large variations in barium concentration locally, between the barium -rich Quad 16 fields, for example, and relatively barium-poor Quad 14 fields. Although sample contamination could be a factor in some individual cases, it is extremely unlikely that it could result in regional variations considering the very different sample types and time spans over which the data were commonly collected and analysed

A simple cross-plot of barium against salinity (Figure 2) shows no obvious relation between the two components: barium is highest in brines of saltiness between 50-100,000 mg/l typically in fields in the central North Sea and Haltenbanken, but even fields with low salinity waters in the northern North Sea such as Magnus and Statfjord contain barium concentrations in excess of 50 mg/l.

These observations are reinforced by statistical analysis of the data (Figure 3) which reveals two main components to the data, the principal component (x-axis trend) is directly related to salinity. The minor component (y-axis trend) is directly related to barium concentration. It appears that the barium-rich formation waters form a statistically distinct population from the majority of North Sea formation waters (but note there is significant complexity within this dataset). Identification of the individual data points in this barium-rich population reveals them to be confined to reservoirs in the late Jurassic Brae formation sandstone (Quad 16 fields). Fields in the northern North Sea also form a population with a strong y-axis component, but it appears that they do not trend to the Brae formation samples, but instead form a separate population.

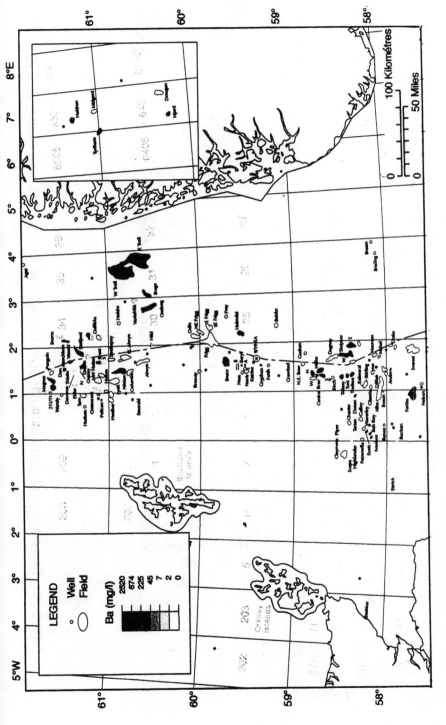

Figure 1: Barium concentration, all data

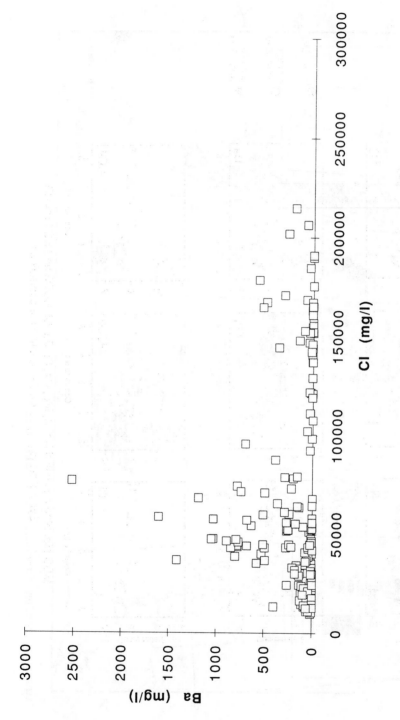

Figure 2: Correlation of barium with chloride for all North Sea formation waters

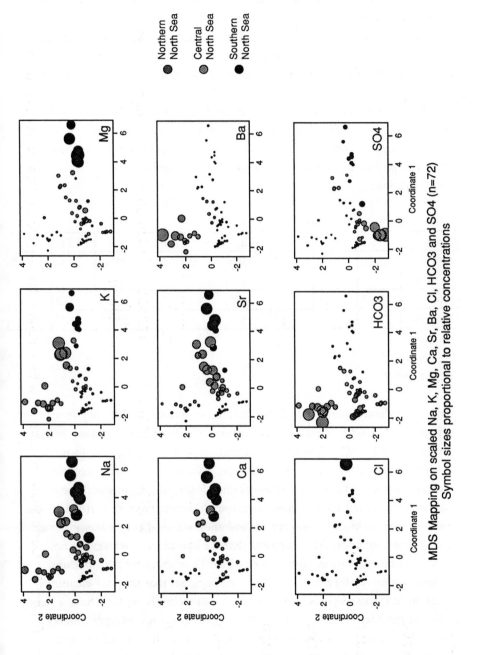

MDS Mapping on scaled Na, K, Mg, Ca, Sr, Ba, Cl, HCO3 and SO4 (n=72)
Symbol sizes proportional to relative concentrations

Further investigation indicates that the barium-rich waters may be distinctive in chemical composition besides simply being barium-rich. Cross-plots of sodium-potassium and sodium-magnesium (Figures 4 & 5)shows that the barium-rich waters have higher potassium and lower magnesium concentrations than barium-poor waters of the similar sodium concentration.

In contrast to barium-rich waters from the Brae formation Quad 16 fields, barium-rich waters in the middle Jurassic Garn formation Haltenbanken fields (north Norway continental shelf) are also strontium-rich. It thus appears that barium-rich waters do not all have identical chemical compositions.

Table 1: Barium-rich formation waters

Field	Na	K	Mg	Ca	Sr	Ba	Cl	HCO3	SO4
Brae Central	44270	1470	73	1380	69	2520	77040	1140	7
Njord	19034	687	380	2400	920	1415	36800	562	
Magnus	13000	230	55	240	50	`117	20650	800	38

DISCUSSION

Barium concentrations in North Sea formation waters do not vary in a systematic way with chemical composition. The barium-rich waters in the central North Sea (Quad 16 fields), all occur in a single geological formation - the late Jurassic Brae formation sandstone. Waters in different geological formations nearby have very different chemical compositions and are not barium-rich. Furthermore, these barium-rich waters have discrete chemical compositions to the barium-poor waters: they are relatively potassium-rich and magnesium poor. This suggests that the barium-rich waters are not simply modified barium-poor waters which have acquired high barium concentrations. Rather, the statistical data suggests that they are chemically different in many respects to barium-poor waters implying they have a very different genetic origin. This is contrary to the previous interpretations of North Sea formation water compositions that they all evolved through mixing of meteoric water and evaporitic brine (Egeberg & Aagaard 1989).

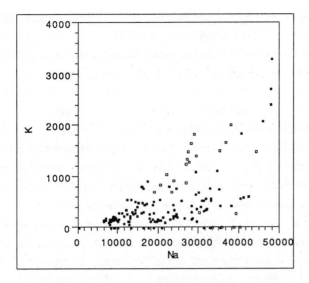

Figure 4
Comparison of K-Na in barium rich (open squares) and barium-poor
formation waters, North Sea

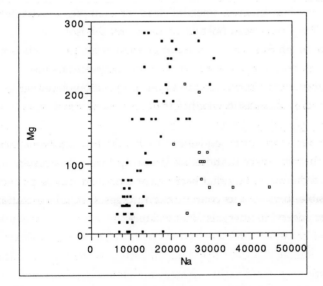

Figure 5:
Comparison of Mg-Na for barium-rich (open squares) and barium-poor
waters, North Sea.

This is further reinforced by the fact that not all barium-rich waters are chemically alike: barium-rich waters from other localities, such as Norway's Haltenbanken, have different major element components. It appears that there is no single common cause of high barium concentration.

Importantly, there is little evidence to suggest that the source of barium is in the reservoirs themselves: the Brae formation sandstones are almost monomineralic, and are composed of quartz (SiO_2). Barium-rich minerals such as potassium-feldspar are virtually absent, which is believed to reflect the provenance of the sands rather than dissolution of pre-existing feldspar grains. Of all the sandstones in the North Sea, these might be expected to contain the least barium as the others do generally contain some K-feldspar. The fact that barium-rich waters form spatially discrete populations, Quad 16 for example, and that their chemical compositions are different in different places suggests that the process controlling their composition is local but outside the reservoir. Shales are a possibility. Although little is known for certain about precise compositional variability in North Sea shales, it is well-established that shales are not uniform. Furthermore, shales contain many chemically reactive constituents, smectite clays for example, which alter during burial. They have been postulated to control the isotopic compositions of North Sea formation waters (Aplin et al. 1993) so it is not unlikely that they may influence major element compositions too. Unfortunately, nothing is known of the shale compositions in either Quad 16 or the Haltenbanken areas as to whether they are also distinctive.

It is clear, however, that North Sea waters are highly heterogeneous, and that the controls on barium concentrations are probably local. As a result, although it is probably unlikely that barium concentration can be predicted blind, it is possible to predict its concentration ranges if local information from equivalent reservoir intervals are available.

CONCLUSIONS

Enormous variations in barium concentrations occur in North Sea formation waters. Statistical analysis discriminates barium-rich waters from all other formation waters: they are distinctive not just in barium content but in high potassium and low magnesium concentrations and are also bicarbonate-rich.

The most barium-rich formation waters are found in Quad 16 fields in the central North Sea; all are confined to the late Jurassic Brae formation reservoir. Barium-rich formation waters in localities outside the North Sea, Norway's Haltenbanken for example, are chemically different. This implies that the controls on barium concentration are both local and different for the two occurrences.

The source of barium does not appear to be within the reservoir itself, but is local. Shales are a possibility. Thus although barium concentration cannot be predicted blind, it is possible to predict it if other data are available from equivalent formations in nearby fields.

REFERENCES

Egeberg, P.K. and Aagaard, P. (1989) Origin and evolution of formation waters from oil fields on the Norwegian shelf. Applied Geochemistry, vol. 4, pp 131-142.

Aplin, A.C., Warren, E.A., Grant, S.M. and Robinson, A.G. (1993). Mechanisms of quartz cementation in North Sea reservoir sandstones: constraints from fluid compositions. AAPG Studies in Geology 36, 7-22.

Warren, E.A. and Smalley, P.C. in press. An atlas of North Sea formation waters. Memoir of the Geological Society of London.

Phosphonate Scale Inhibitor Adsorption on Outcrop and Reservoir Rock Substrates – The 'Static' and 'Dynamic' Adsorption Isotherms

M. D. Yuan, K. S. Sorbie, P. Jiang, P. Chen, M. M. Jordan, and A. C. Todd

DEPARTMENT OF PETROLEUM ENGINEERING, HERIOT WATT UNIVERSITY, EDINBURGH, EH14 4AS, UK

K. E. Hourston

TOTAL OIL MARINE, ABERDEEN, UK

K. Ramstad

NORSK HYDRO, BERGEN, NORWAY

1 ABSTRACT

Chemical scale inhibitors, applied in field "squeeze" treatments, provide the main approach for the prevention of sulphate and carbonate downhole scale formation in oil reservoirs. In adsorption squeeze treatments, the inhibitor/rock interaction which governs the dynamics of the inhibitor return profile is described by an adsorption isotherm which may be a function of inhibitor concentration, pH, temperature, $[Ca^{2+}]$, etc; that is, $\Gamma = \Gamma(C, [H^+], T, [Ca^{2+}], ..)$. The nature of the isotherm also depends strongly on the adsorbing mineral substrate (e.g. quartz, clays, feldspar,..) and the inhibitor type (e.g. phosphonate, phosphinocarboxylate, polyacrylate etc.). The form of the adsorption isotherm and its sensitivities to these various factors may be investigated using either (i) relatively simple static beaker tests using quartz, clay mineral separates or crushed core material; or (ii) more complex dynamic core flooding experiments. This paper presents a study of the adsorption characteristics of the phosphonate scale inhibitors in both static tests and in dynamic flow conditions. The uses and limitations of each type of experiment are explained and a number of detailed results are presented.

A number of adsorption core floods have been carried out with phosphonate scale inhibitors at both ambient (20°C) and elevated (>100°C) temperatures. Several North Sea reservoir rocks were used as core material and some core floods in outcrop sandstone were used for comparative purposes. The inhibitor "dynamic" adsorption isotherms were then derived using the inhibitor effluent concentration data obtained from these flooding experiments. Supporting static beaker tests of phosphonate adsorption on the crushed core material have also been carried out to measure the "static" adsorption isotherms for comparison with the "dynamic" isotherms.

It is the dynamic isotherms, derived from the inhibitor core floods, that are relevant for analysing and designing field inhibitor squeeze treatments. These capture the appropriate interactions between the inhibitor, solvent and rock substrate for modelling purposes. Of particular importance is the slope of the isotherm, $(d\Gamma/dC)$, in the threshold inhibitor concentration region, since this determines the squeeze lifetime. However, the static adsorption isotherms can be used to establish the overall inhibitor adsorption *sensitivities* (e.g. effects of temperature, pH, $[Ca^{2+}]$ and clay minerals) and this is a very useful feature of such experiments. However, the static tests cannot be used for establishing the precise magnitude of the inhibitor adsorption or for finding the shape of the isotherm curves. In particular, the static adsorption isotherms do not possess sufficient vital information in the threshold inhibitor concentration region. An example of where the complementary nature of static and dynamic adsorption tests has been used is in establishing the effect of the presence of certain clay minerals on inhibitor adsorption isotherms, which is discussed in this paper.

The inhibitor return profiles from both inhibitor core floods and squeeze treatments are then analysed using the corresponding dynamic adsorption isotherms. This demonstrates how the inhibitor returns are related to the shape of the dynamic isotherms. Our approach to using appropriately collected core flood data for the design of squeeze treatments is illustrated through two field cases where the optimisation of the "Field Squeeze Strategy" is illustrated. We show that, for each field case, a different operational factor is involved in this optimisation process.

2 INTRODUCTION

Phosphonates are one of the main types of scale inhibitors which are used in downhole squeeze treatments to prevent scale deposition in producer wells (Vetter, 1973; Meyers, *et al*, 1985; King and Warden, 1989; Kan, *et al*, 1991; Breen, *et al*, 1991). The retention of phosphonate molecules in a reservoir formation is normally controlled by an adsorption/desorption process in which the inhibitor return curve is mainly governed by the following factors (Sorbie, 1991; Sorbie, *et al*, 1991a, 1992):

a) the shape of the inhibitor adsorption isotherm, $\Gamma(C)$, where the most important factor is the steepness of the isotherm at inhibition threshold concentrations i.e. $(d\Gamma/dC) \gg 1$;

b) whether or not the inhibitor adsorption isotherm shows a plateau behaviour at higher concentrations; and

c) to a lesser extent, the nature of the non-equilibrium adsorption which may also have an effect on the inhibitor return profile.

The adsorption/desorption behaviour of a phosphonate inhibitor in porous media is usually evaluated by conducting static bulk adsorption tests (i.e. beaker tests) and/or inhibitor dynamic core flooding experiments (Vetter, 1973; Meyers, *et al*, 1985; King and Warden, 1989; Kan, *et al* 1991, 1992; Breen, *et al*, 1991; Sorbie, *et al*, 1992, 1993a, 1993b). Static adsorption isotherms from beaker tests and inhibitor core flood return curves have been conventionally relied upon to indicate how well an inhibitor adsorbs and desorbs on a given rock material. However, this gives rise to the question of whether such static adsorption isotherms and core flood return curves can be used to determine the inhibitor performance in a field squeeze treatment and, if it can, how? Recently, a methodology was proposed to derive inhibitor dynamic adsorption isotherms using inhibitor core flood effluent concentration data and this can be applied to develop the "Field Squeeze Strategy" (Sorbie, *et al*, 1991a, 1992; Sorbie and Yuan, 1993; Yuan *et al*, 1993) .

In this paper, we present results from some of the inhibitor adsorption core floods which have been carried out with the phosphonate scale inhibitors at both ambient (20°C) and elevated (>100°C) temperatures. Several North Sea reservoir rocks were used as core material and some core floods in outcrop material (Clashach sandstone) were also used for comparative purposes. The inhibitor dynamic adsorption isotherms were then derived using the inhibitor effluent concentration data obtained from these flooding experiments. Supporting static beaker tests on phosphonate adsorption on the crushed core material were also conducted to measure the static adsorption isotherms.

The corresponding static and dynamic adsorption isotherms are compared and their respective uses are discussed. The dynamic adsorption isotherms are then used to analyse the inhibitor core flood effluent concentration profiles and squeeze inhibitor returns. The effect of the presence of clay minerals on the inhibitor return profiles and the shape of the isotherms is also discussed.

In order to extrapolate phosphonate inhibitor adsorption performance on a laboratory scale to field squeeze treatments, two of the dynamic adsorption isotherms derived from reservoir condition field core flood data were input to our software, SQUEEZE (Sorbie and Yuan, 1993) to predict the lifetimes of two planned field treatments. Modelling results show that, if the phosphonates perform the same in the field as they did in the core, both squeeze treatments would give satisfactory inhibitor returns and squeeze lifetimes. Furthermore, based on the modelling using the isotherm data, the design operational parameters (i.e. inhibitor concentration, inhibitor volume and overflush volume) for both treatments are shown to have a scope for improvement from the original designs in order to extend the squeeze lifetimes. We had previously demonstrated the usefulness of computer modelling for history-matching field squeeze returns of both adsorption and "precipitation" types and for improve inhibitor placement strategy in heterogeneous reservoir formations (Yuan, *et al*, 1993). In this paper, we again illustrate the usefulness of computer modelling for "extrapolating" scale inhibitor retention performance on a laboratory scale to field squeeze treatments.

3 SUMMARY OF THE EXPERIMENTS

The experiments are described only in brief outline here since the intention of this paper is to illustrate the theoretical and quantitative approach by which the scale inhibitor adsorption/desorption behaviour can be evaluated and translated to field squeeze treatments.

Inhibitors: The phosphonates used were supplied by three different companies and are labelled as I1, I1a and I1b respectively. The activity of I1 is ~ 50% and the activities of I1a and I1b are both ~ 29%. Throughout this paper, inhibitor adsorption is quoted in mg inhibitor per gram of rock grains and concentration is in ppm and, unless otherwise specified, both are based on the *active* content of the supplied inhibitor solutions.

Rock materials: Both the highly quartzitic Clashach, an outcrop sandstone quarried in the Elgin region of Scotland, and the reservoir core materials drilled from Brent group formations in North Sea, were used for static adsorption tests (crushed to particles) and dynamic flow experiments (consolidated).

Brines: Synthetic North Sea seawater was used for the preparation of inhibitor solutions at a required concentration and for brine preflush and postflush of a core. The main ion components of the seawater are 2960 ppm $[SO_4^{2-}]$ and 428 ppm $[Ca^{2+}]$. The full composition has been presented previously (Sorbie, *et al*, 1993a, 1993b).

Static adsorption tests: 15 ml of inhibitor seawater solution of required concentration and pH were added to 15 grams of crushed core material particles of size ranging from 32 to 600μm (surface area is 0.93 m^2/g) in a capped plastic bottle. The bottle was placed in a thermo-static bath and shaken regularly for 6 hours, left undisturbed for 12 hours and then shaken hourly for 6 hours before being filtered through a 0.45μm paper filter. The sample was then analysed for inhibitor concentration in the solution and adsorption level was calculated.

Dynamic core floods: All core floods are summarised in **Table 1**. Two floods (floods F1 and F2) were carried out at ambient conditions while the others were reservoir conditioned core floods (floods F3 to F8, with residual oil saturation and at reservoir temperatures). The flow rate used for inhibitor injection and seawater postflush was 30 ml/hour for all the floods. For a typical flood, a 1" x 6" core was placed in a Hassler type core holder. A confining pressure of 1000 psi was applied to the rubber sleeve in the core holder and a back pressure of 80 psi was also applied for reservoir temperature floods. The cores used for reservoir condition floods (i.e. F3 to F8) were solvent refluxed and residual oil saturated with dead crudes. A typical flood sequence included:

i)   Injection of a spearhead of surfactant solution at ambient temperature (20˚C) was carried out for floods F3 - F6. No spearhead was injected in floods F1, F2, F7 and F8.

ii)   Injection of phosphonate seawater solution was carried out at the required concentration and pH at ambient temperature (20˚C) until the effluent inhibitor concentration had reached the input level. This usually required an inhibitor slug of 4 to 8 pore volumes.

Table 1   SUMMARY OF EXPERIMENTAL DETAILS OF PHOSPHONATE SCALE INHIBITOR CORE FLOODS

Flood Information	Flood F1	Flood F2	Flood F3	Flood F4	Flood F5	Flood F6	Flood F7	Flood F8
Core material	Clashach outcrop	Clashach outcrop	field core (Tarbert)	field core (Tarbert)	Clashach outcrop	Clashach outcrop	field core (Oseberg)	field core (Etive)
Phosphonate Inhibitor Types	I1	I1	I1a	I1b	I1a	I1a	I1b	I1b
Concentration injected (active)	2500ppm	2500ppm	57952ppm	43181ppm	58881ppm	60077ppm	40818ppm	42492ppm
Inhibitor slug pH	4.0	6.0	2.0	2.4	2.0	2.0	2.4	2.4
Temperature for preflush and inh. injection	20°C	20°C	20°C	20°C	20°C	20°C	20°C	20°C
Temperature for shut-in and postflush	20°C	20°C	110°C	110°C	110°C	110°C	106°C	106°C
Any residual oil sturation in core?	no	no	yes	yes	yes	yes	yes	yes
Any demulsifier/ surfactant preflush ?	no	no	yes	yes	yes	yes	no	no
Type of brine used for postflush	synthetic seawater	synthetic seawater	synthetic seawater	synthetic seawater	synthetic seawater	synthetic seawater (Ca and Mg removed)	synthetic seawater	synthetic seawater

Table 2   INHIBITOR ADSORPTION AND RECOVERY CALCULATIONS FOR THE RESERVOIR CONDITION CORE FLOODS

Flood No.	Recovery, 1 (%)	Recovery, 2 (%)	Γ (mg/g)
3 (Tarbert core)	96.29	76.00	3.425
4 (Tarbert core)	96.97	79.96	2.521
5 (Clashach core)	96.31	76.50	1.982
6 (Clashach core)	96.21	68.41	1.962
7 (Oseberg core)	96.13	73.19	2.920
8 (Etive core)	95.47	68.25	3.528

Recovery, 1 - percent of *total injected inhibitor mass* being recovered at end of experiment.
Recovery, 2 - percent of *inhibitor mass in core before postflush* being recovered at end of experiment.
Γ - inhibitor adsorption level in core before seawater postflush (after adsorption shut-in).

iii)   Flow was stopped and temperature was raised to the reservoir temperature (>100°C) and the core was left for a 20 hour adsorption shut-in.

iv)   The core was postflushed with synthetic seawater (at reservoir temperature) until the effluent inhibitor concentration had fallen below the target inhibition level (< 5ppm).

Note that floods F1 and F2 were conducted at ambient temperature throughout the flood cycle.

Inhibitor assay: The phosphonate inhibitor concentrations were analysed using the persulphate UV oxidation method.

4 STATIC AND DYNAMIC ADSORPTION ISOTHERMS

Phosphonate adsorption isotherms were measured in static bulk solution adsorption tests and were also derived from the dynamic inhibitor core flood effluent concentration data. Note that, for comparative studies, the static adsorption tests were carried out with the same inhibitor, solution pH, temperature and core material (crushed) as those used for corresponding consolidated core inhibitor floods.

         Figure 1 and **Figure 2** show two examples of comparisons between the static and dynamic adsorption isotherms for crushed or consolidated Clashach sandstone which is a highly quartzitic substrate. The only difference between these two figures is that they are carried out at different pH values of 4 and 6 for **Figures 1** and **2**, respectively. The static adsorption isotherms show clear differences from the dynamic isotherms in both the magnitude of the adsorption levels and the shape of the isotherms. First, the amount of maximum adsorption at full concentration (i.e. 2500 ppm) measured from a static adsorption test is 2-4 times that derived from a comparable inhibitor core flood. Secondly, the static isotherm curves do not rise very sharply at low concentration and then level out, which is the case for the dynamic isotherms. Several factors are believed to cause such differences:

i.   Crushing of core material exposes fresh rock surface which does not represent the true surface properties of rock pores and also increases specific surface area of the rock substrate.

ii.   The relatively low liquid/solid ratio present in a core flood cannot be reproduced in static adsorption tests since a sufficiently large volume of solution sample is required for inhibitor analysis.

iii.   The transport of inhibitor molecules from bulk solution onto rock surface may be restricted in static tests in contrast to core flow experiments, although rotating bottles may alleviate this problem.

iv.   The adsorption/desorption equilibrium at static conditions (a higher adsorption) may differ from that under a dynamic flowing conditions (a lower adsorption) (Sorbie et al, 1993a).

         Because of the above differences, and also because it is extremely difficult to accurately measure inhibitor adsorption levels in static adsorption tests in the very low inhibition threshold concentration region (< 10 ppm), it is inappropriate to use static isotherms to analyse an inhibitor core flood and to predict a squeeze lifetime. For example, if the pointwise value of $\Gamma(C)$ is not accurate, then the more important quantity, $(d\Gamma/dC)$, will be even more in error (Sorbie et al, 1991a). However, static adsorption tests are very quick compared with inhibitor core flooding experiments and can be very useful to establish the overall of inhibitor adsorption **sensitivities** to say $[Ca^{2+}]$, pH, temperature and clay minerals etc. **Figures 1** and **2**, for example, show both the static and dynamic isotherms for pH 4 and 6 and, although these isotherms are different, the static isotherm sensitivity to pH is the same as the dynamic isotherm i.e. higher phosphonate adsorption onto quartz at 20°C at pH 6 than for pH 4 at C = 2500 ppm. Likewise, **Figure 3** shows an example showing more details of

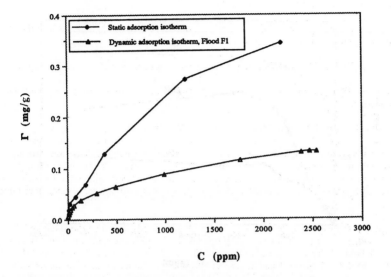

Figure 1   Comparison between a static adsorption isotherm and a dynamic adsorption isotherm. Inhibitor solution pH 4 and ambient temperature

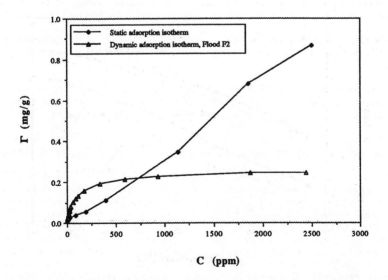

Figure 2   Comparison between a static adsorption isotherm and a dynamic adsorption isotherm. Inhibitor solution pH 6 and ambient temperature

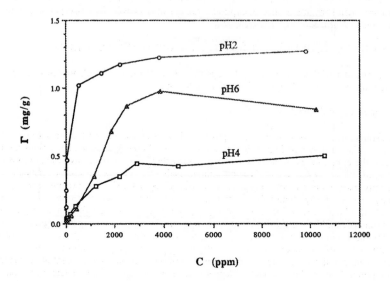

Figure 3   Phosphonate static adsorption isotherms at pH 2, 4 and 6 and 25°C

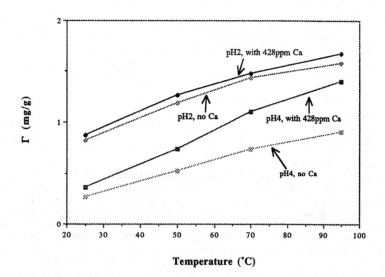

Figure 4   Effects of calcium ions and temperature on phosphonate static adsorption

the pH sensitivity of phosphonate inhibitor adsorption and **Figure 4** shows adsorption sensitivity to temperature and Ca^{2+}, both measured from static bulk solution tests. These trends with pH and temperature, seen here in the bulk adsorption tests, are also seen in the corresponding core floods (Sorbie, *et al*, 1993b).

5 CHARACTERISTICS OF THE DYNAMIC ADSORPTION ISOTHERMS

For the remainder of this paper, we will focus mainly on the dynamic adsorption isotherms since these are more relevant for designing field squeeze treatments.

The Shape of the Isotherms

The phosphonate inhibitor dynamic adsorption isotherm curves derived from reservoir condition floods F3 to F8 are shown in **Figure 5**. The isotherms in **Figure 5** were validated by using them to reproduce the experimental return curves from which they were derived, as discussed previously (Sorbie *et al*, 1991a, 1992). All of these phosphonate adsorption isotherms characteristically show a very steeply rising section at low concentrations ($(d\Gamma/dC) \gg 1$ for $C < 50\text{-}100$ ppm) and a gradually flattening section (near-plateau behaviour) at higher concentrations ($C > 200$ ppm). As discussed previously (Sorbie, 1991, Sorbie, *et al*, 1991a, 1992), this shape of the isotherm is very favourable for an extended inhibitor return if the inhibitor is applied to a field adsorption squeeze treatment. The dramatic levelling-off at $C > \sim 200$ ppm (this concentration is still very low compared with a commonly used 15% supplied inhibitor concentration or $\sim 5\%$ active concentration) suggests there is a sudden change of the surface adsorption of inhibitor molecules once the concentration has reached a certain point. The combined effect of deeper placement (because of the plateau behaviour of the isotherm) and retarded return velocity would then produce an extended squeeze return before the residual concentration falls below the inhibition threshold concentration. This adsorption behaviour will be discussed further in relation to inhibitor return profiles later in this paper.

Effect of Clay Minerals on Phosphonate Adsorption

The reservoir condition core floods were carried out using two broad types of rock materials. The Clashach outcrop sandstone used for floods F1, F2, F5 and F6 is highly quartzitic (> 95% quartz content) with 3-4% feldspar and mica and less than 1% clay content. On the other hand, the reservoir cores from Brent group formations have only $\sim 80\%$ quartz but $\sim 10\%$ clay content (predominantly kaolinite). Because of both the different surface properties of kaolinite and quartz and the significantly higher specific surface area of kaolinite compared with quartz, it is interesting to compare the adsorption performances of phosphonates on the two rock substrates.

Figure 5 shows that the adsorption levels of phosphonate in the reservoir cores (F3, F4, F7 and F8) are generally higher than those in Clashach cores (F5 and F6). This is confirmed from the inhibitor material balance calculation carried out for these floods which is given in **Table 2** and shows that the inhibitor adsorption on the reservoir cores after the shut-in is between 2.5 to 3.5 mg/g while that in the Clashach core is less than 2.0 mg/g. This difference is believed to be caused by the high clay content in the reservoir cores. Due to the higher adsorption and generally steeper isotherms at medium and high inhibitor concentrations as seen in **Figure 5**, the same inhibitor would produce a higher concentration return from reservoir rocks in such concentration regions. However, because the isotherms obtained using both core materials are very similar at very low concentration range (< 10 ppm), the inhibitor squeeze lifetimes from both Clashach and reservoir rocks would be very close for say a 5 ppm threshold inhibition level. In other words, the effect of clay mineral presence in reservoir formation would be important for a squeeze lifetime if the formation water has a severe scaling tendency and the inhibition threshold concentration is relatively high (> 10 ppm), but it is less important if the water scaling tendency is moderate and the inhibition threshold is low (say, < 5 ppm). A fuller description of the effect of clay minerals on phosphonate adsorption has been presented previously (Jordan, *et al*, 1994a).

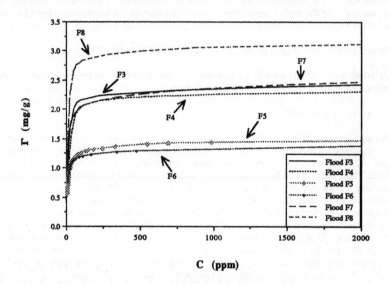

Figure 5   Phosphonate inhibitor dynamic adsorption isotherms derived from reservoir condition core flood data

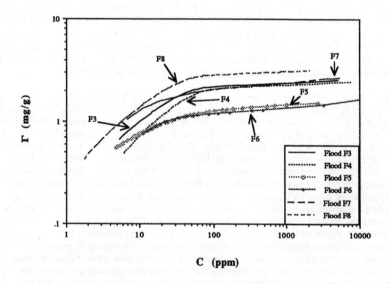

Figure 6   Phosphonate inhibitor dynamic adsorption isotherms derived from reservoir condition core flood data. Log plots

Apart from the effect on scale inhibitor retention, the presence of clay minerals in the reservoir rocks may also contribute to formation damage during a squeeze treatment (Jordan, *et al*, 1994b). This should be taken into consideration when choosing inhibitor solution pH and spearhead surfactant for a squeeze treatment but further discussion of this is beyond the scope of this paper.

Adsorption Isotherms: Comparison with Freundlich and Langmuir Forms

Freundlich and Langmuir adsorption isotherms are commonly used to describe scale inhibitor adsorption on rock surfaces (Hong and Shuler, 1988; Sorbie, *et al*, 1991b, Yuan, *et al*, 1993; Sorbie and Yuan, 1993; Shuler, 1993). Both types of isotherm are expressed by a two parameter equation as follows:

a. Freundlich isotherm equation:

$$\Gamma_{eq}(C) = kC^n \tag{1}$$

b. Langmuir isotherm equation:

$$\Gamma_{eq}(C) = \frac{abC}{1+bC} \tag{2}$$

where, Γ_{eq} is the equilibrium adsorption level (mg per g of rock or mg per litre of rock grains) at concentration, C; k and n are Freundlich parameters and a and b are Langmuir parameters, respectively.

We have used both equations for describing inhibitor adsorption in our modelling of both field squeeze inhibitor returns and laboratory core flood inhibitor returns and proved that for most cases the measured inhibitor return data can be modelled with either of the equations. It is interesting, then, to see whether the isotherm data which is actually derived from our core flooding experiments resemble either of the types of adsorption behaviour. This has been examined simply by plotting inhibitor adsorption levels versus concentrations in two different ways as described in the following.

Taking the logarithm of each side of Eq.(1) gives:

$$\log\left[\Gamma_{eq}(C)\right] = K + n\log(C) \tag{3}$$

where $K = \log(k)$, and is a constant. Eq.(3) shows that the log of the adsorption level, $\log(\Gamma_{eq})$, has a linear relation with the log of the concentration, $\log(C)$, for Freundlich type of adsorption.

For the Langmuir form in Eq. (2), the reciprocal of the adsorption level, $(1/\Gamma_{eq})$, has a linear relation with the reciprocal of the concentration, $(1/C)$, as follows:

$$\left[\frac{1}{\Gamma_{eq}(C)}\right] = \frac{1}{a} + \frac{1}{ab}\left(\frac{1}{C}\right) \tag{4}$$

With Eqs.(3) and (4) in mind, the experimentally derived isotherm data were then plotted logarithmically, as shown in **Figure 6**, and in inverted form, as illustrated in **Figure 7**. Clearly the data points of each isotherm are not on a straight line for either form of plotting. This indicates that the dynamic isotherms deviate from both Freundlich type of adsorption and Langmuir type of adsorption.

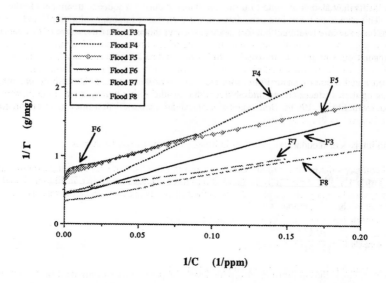

Figure 7   **Phosphonate inhibitor dynamic adsorption isotherms derived from reservoir condition core flood data. Inverted plots**

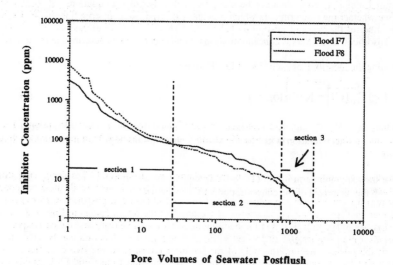

Pore Volumes of Seawater Postflush

Figure 8   **Phosphonate inhibitor effluent concentration profiles during floods F7 and F8.**

6 ANALYSIS OF INHIBITOR RETURN PROFILES USING THE DYNAMIC ISOTHERMS

It has often been pointed out that the inhibitor return curve (squeeze lifetime) is mainly governed by the shape of the adsorption isotherm, although the rate of adsorption/desorption may also have some effect, especially during the injection and shut-in stage (Sorbie, 1991; Sorbie, *et al*, 1991a, 1992). In this section, this view is illustrated clearly by: (a) comparing a pair of core flood inhibitor return profiles (i.e. from floods F7 and F8) within the context of the shape of the adsorption isotherms derived using these core flood inhibitor effluent data; and (b) simulating squeeze inhibitor returns with two isotherms - one the original from flood F5 and the other modified at the higher concentration section of the isotherm.

Inhibitor Core Flood Return Profiles in Relation to Adsorption Isotherms

Figure 8 compares the inhibitor return curves measured from the two core flooding experiments F7 and F8 (**Table 1**). The curves may be divided into three sections, as shown in **Figure 8**, and each section may then be analysed in connection with their respective adsorption isotherms. That is:

• **section 1** is from 0 to 25 pv of seawater postflush;

• **section 2** is from 25 to 750 pv;

• **section 3** is from 750 pv to the end of experiments.

The first section, i.e. the first 25 pore volumes of postflush and C > ~ 100 ppm, shows that the effluent inhibitor concentration from flood F7 is higher than that observed for flood F8. In other words, the inhibitor return from F7 is more retarded than that from F8, or, more pore volumes of seawater postflush were required for flood F7 than for F8 in order to reach the same effluent concentration. Now, let us consider this together with the shape of the adsorption isotherms at the corresponding concentration region (i.e. C > 100 ppm, see **Figure 9**). Note that inhibitor return velocity, v_c, at a given concentration, C, is inversely proportional to the derivative $(d\Gamma/dC)$ of the adsorption isotherm at C, which is given by:

$$v_c = \frac{v_w}{1 + \frac{1-\phi}{\phi}\left(\frac{\partial\Gamma}{\partial C}\right)_c} \tag{5}$$

where v_w is the water velocity and ϕ is porosity. In Eq. (5), the units of adsorption (Γ) are in mass per unit volume of *rock grain* (as in Hong and Shuler, 1988) and C is in mass per unit volume of solution.

As **Figure 9** shows, at C > 100 ppm, the isotherm curve from flood F7 is rising more steeply than that from flood F8 (although the adsorption *level* in flood F7 is lower), while the corresponding inhibitor return from F7 over the same region (i.e. C > 100 ppm and postflush < 25 pv) is more retarded than that from flood F8, as shown in **section 1** of the return curves in **Figure 8**. This is exactly what we expect from the relation between inhibitor return velocity and the derivative of the inhibitor adsorption isotherm as expressed in Eq.(5). A similar explanation can be given for the profiles of the lower concentration tails of the inhibitor return from the two core floods i.e. in **sections 2** and **3** of Figure 8. The inhibitor return concentration from flood F8 is higher (more retarded) in the region between 25 to 750 pv postflush (C = ~ 10 ppm - 100 ppm) than that from flood F7 (see **Figure 8**, section 2 of the return curves). This correlates very well with the shape of the isotherms in the same concentration region as shown in **Figure 10**, where the flood F8 isotherm is steeper than the flood F7 isotherm (the adsorption *level* in flood F8 is also higher than F7, in this

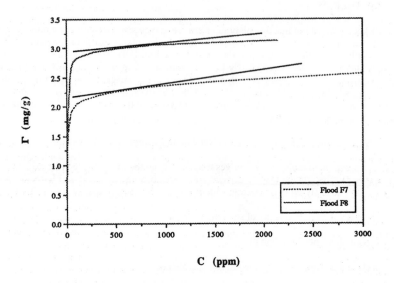

Figure 9   Phosphonate inhibitor dynamic adsorption isotherms derived from flood F7 and F8 effluent data

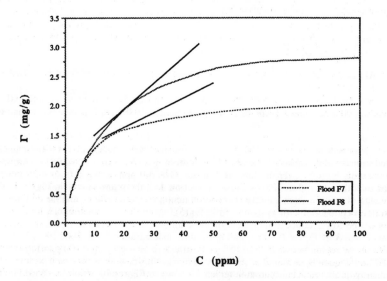

Figure 10   Phosphonate inhibitor dynamic adsorption isotherms derived from flood F7 and F8 effluent data.   0-100ppm region

case). In the long tail region close to the threshold inhibitor concentrations (C < 10 ppm and postflush > ~ 750 pv), **Figure 11** shows both the adsorption levels and the derivatives of the isotherms from the core floods are converging together. This results in very similar inhibitor return curves for floods F7 and F8 (see **Figure 8**, section 3 of the return curves) in this low concentration region.

Thus, all sections of the inhibitor return curves can be clearly explained with reference to the shape (i.e. the *slope*) of the corresponding inhibitor adsorption isotherms. This is illustrated here using the results from two field core floods.

Squeeze Inhibitor Return Profiles in Relation to Adsorption Isotherms

In order to further demonstrate the correlation between an inhibitor return with the shape of its adsorption isotherm, modelling of a field squeeze treatment was carried out. The application strategy for this treatment was taken from that of a field operation which was subsequently carried out. Two dynamic adsorption isotherms were used separately as input to the model in order to simulate inhibitor returns. While *Isotherm 1* was the data actually derived from core flood F5 (see **Table 1**), *Isotherm 2*, was modified from the F5 isotherm by deliberately raising the inhibitor adsorption levels (and hence the derivatives) at concentration above 10 ppm while preserving the critical threshold concentration section (C =< 10 ppm). **Figure 12** compares the two isotherms which, as noted, are identical below 10ppm. The purpose of this modelling exercise using *Isotherms 1* and 2 is to illustrate that an adsorption squeeze lifetime is mainly governed by the shape of the adsorption isotherm in the region of the inhibition threshold concentration, even though the return profile during early production stage may be affected by the isotherm shape at higher concentrations.

Figure 13 presents the two inhibitor residual concentration profiles simulated with the identical squeeze treatment design using *Isotherms 1* and 2. Clearly, return curve 2 is higher (more retarded) during first 75 days than return curve 1 because *Isotherm 2* (modified) is more steeply rising than *Isotherm 1* (original) in the corresponding concentration range (see **Figure 12**). However, there is little difference in the very tail region (i.e. 75 - 180 days or 5 - 10 ppm) between the two return curves. In fact, we would argue this difference is hardly detectable in real produced brines because of a range of sampling and analytical limitations (see discussion in Graham *et al*, 1993). The similar squeeze lifetimes produced from the two isotherms (at 5 ppm say), illustrates and confirms our view that it is the shape in this region of the isotherm which dominates the dynamics of the inhibitor return curve for an adsorption squeeze process.

7 APPLICATION OF DYNAMIC ADSORPTION ISOTHERMS TO FIELD SQUEEZE TREATMENTS

Field Modelling Approach and Limitations

The adsorption isotherms from flood F3 and flood F8 (see **Table 1**), represented as a table of data, were used in our software (SQUEEZE; Sorbie and Yuan, 1993) to model certain field applications. Case studies predicting treatment lifetimes for two field squeezes were carried out and results were used in the optimisation of squeeze treatment design. The purpose of this exercise was to apply the inhibitor adsorption/desorption results, as observed in the reservoir cores, to field squeeze performance of the inhibitor. In doing so we must, of course, assume that the inhibitor would perform in a very similar way in the reservoir formation as it did in the inhibitor core floods in the laboratory.

Before proceeding, it is helpful to recall the limitations of using modelling to predict field behaviour in the context of inhibitor squeeze treatments. In the modelling results presented below, we frequently quote predicted squeeze lifetimes for a required threshold inhibitor concentration, C_t. However, we note that the modelling is certainly *not* reliable enough to quote squeeze lifetimes accurately. The main uses of the SQUEEZE modelling approach are primarily as follows:

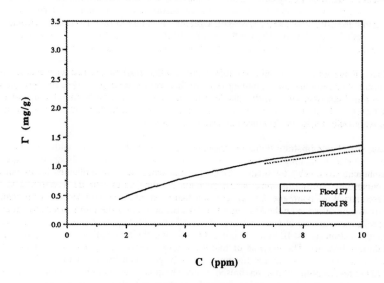

Figure 11   Phosphonate inhibitor dynamic adsorption isotherms derived from flood F7 and F8 effluent data. 0-10ppm region

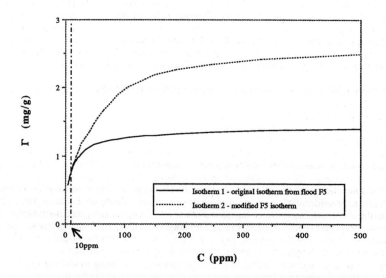

Figure 12   Comparison between the original adsorption isotherm derived from flood F5 and the modified curve (steeper at C >10ppm)

(i) to establish whether the squeeze lifetime for the measured isotherm is "satisfactory";

(ii) to examine **sensitivities** to factors such as inhibitor concentration and overflush size etc.

(iii) to use these sensitivities in order to design an optimised "Field Squeeze Strategy".

These modelling aims should be kept in mind when reading the results presented below. Further assumptions in our modelling approach are that we are treating a single reservoir sand and that the core is "representative" of this single reservoir zone. When more complex heterogeneous reservoir facies are being treated, this requires a different approach to the modelling as discussed in detail elsewhere (Sorbie, *et al,* 1991b, Yuan *et al,* 1991, 1993).

Base Case Scale Inhibitor Squeeze Strategies

The phosphonate scale inhibitor squeeze treatments presented here were planned for two different fields in North Sea. The adsorption isotherm data used for the studies were derived from core floods which were carried out using the core materials drilled from the respective formations in which the squeeze treatments were to be carried out. The original designs of the two squeeze treatments are shown as follows:

Original design for squeeze treatment 1 in field A (Original Design 1):

*   *100 bbls seawater preflush*

*   *1215 bbls 20% inhibitor solution (~ 5.8% active content)*

*   *followed by 1618 bbls seawater overflush*

*   *12 hour shut-in*

*   *start back production and monitor inhibitor residuals*

Original design for squeeze treatment 2 in field B (Original Design 2):

*   *50 bbls seawater preflush with 0.06% phosphonate (0.02% active content)*

*   *67 bbls 15% phosphonate inhibitor solution (~ 4.4% active content)*

*   *followed by 790 bbls seawater overflush*

*   *22 hour 40 minutes shut-in*

*   *start back production and monitor inhibitor residuals*

The adsorption isotherm data from the respective core floods (F3 for Squeeze Treatment 1 and F8 for Squeeze Treatment 2) were then input to our software in order to predict the squeeze lifetimes based on the above original squeeze designs. Following this stage, an optimisation of the squeeze operational parameters (inhibitor concentration, inhibitor slug volume and overflush size) was carried out in order to improve on the squeeze lifetimes as predicted based on the original operations. The results from this modelling study of squeeze treatment designs is presented below.

Case Study 1 - Squeeze Treatment 1 in Field A

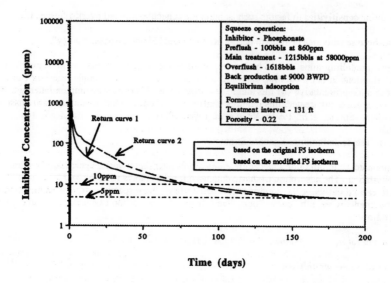

Figure 13   **Comparison of simulated squeeze inhibitor returns using the original and the modified flood F5 adsorption isotherms (see Figure 12)**

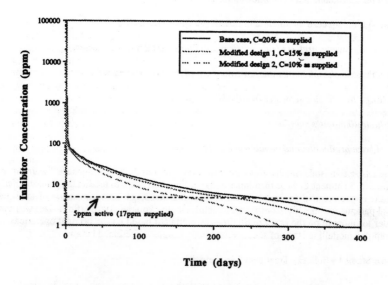

Figure 14   **Comparison of squeeze inhibitor returns simulated with the Original Design 1 and the modified designs altering inhibitor concentration**

For the Original Design 1, the predicted squeeze lifetimes from SQUEEZE using the isotherm data from flood F3 are very satisfactory as presented in **Table 3**. They are predicted to be 137 days at 10 ppm or 258 days at 5 ppm, noting the qualification on such predicted squeeze lifetimes discussed above. Variations from the base case in inhibitor injection concentration and seawater overflush volume were then modelled. These results are illustrated in **Figure 14** for the concentration sensitivity and in **Figure 15** for overflush sensitivity and the corresponding squeeze lifetimes under these varied operational conditions are compared with the base case in **Table 3**.

The results in **Figure 14** suggest that 15% (~ 4.4% active) inhibitor concentration would give a reasonable squeeze lifetime compared with the 20% (~ 5.8% active) for the Original Design, although a 10% solution (~2.9% active) is perhaps rather too low. The sensitivities in **Figure 15** indicate that a larger overflush volume than that planned would increase the squeeze lifetime quite significantly. It is apparent that, for this particular case, the planned squeeze treatment is sensitive to overflush size at least as much as to inhibitor concentration. Indeed, this particular recommendation was implemented in the field and the evidence strongly suggests that the appropriately increased overflush results in an extended squeeze lifetime from previous treatments (which resembled the Original Design 1 presented here). This case will be reported in more detail elsewhere (Jordan *et al*, 1994c).

Case Study 2 - Squeeze Treatment 2 in Field B

In Case Study 2, a much smaller inhibitor treatment was envisaged in the base case application followed by a relatively larger overflush. **Table 4** lists the squeeze lifetimes predicted for threshold active concentrations of 10 ppm, 5 ppm and 1.4 ppm where we have used the isotherm derived from flood F8 (Table 1). This core was again from the reservoir sand for which the squeeze was intended. In this case, had the inhibitor performed in the field in the same way as in core flood F8, the squeeze lifetime would be quite satisfactory for the well where a squeeze lifetime of ~ 160 days at 1.4 ppm (5 ppm product as supplied) is predicted. A number of sensitivity calculations were then performed using SQUEEZE in order to investigate the effects of inhibitor concentration, inhibitor main treatment volume and overflush volume, etc on the predicted squeeze lifetime compared with the calculated base case lifetime (i.e. Original Design 2).

From the sensitivity study, it was found that the squeeze treatment can be optimised with changes to the main treatment volume and overflush volume. Compared with the Original Design 2, an optimised design was found in which the main inhibitor treatment volume is increased from 67 bbls to 268 bbls (where inhibitor concentration is unchanged) and the seawater overflush volume is halved to 395 bbls. The simulated inhibitor return from the altered design is compared with that simulated for the base case treatment in **Figure 16**. **Table 4** also summarises the squeeze lifetimes predicted based for both the Original Design and the modified treatment design. Using the isotherm from flood F8, the squeeze lifetime at 1.4 ppm (5 ppm product) is predicted to be 336 days with the new design compared with 163 days for the base case design. For this particular case, while the main factor influencing inhibitor squeeze return is found to be the inhibitor slug volume, the recommended reduction in overflush size results from the concern about the potential problems with well lift after the squeeze operation and from the fact that the effect of overflush on the squeeze lifetime is insignificant for this treatment.

The two cases above serve to illustrate that the sensitivities of different squeeze treatments to various operational parameters may be quite different. Bearing this in mind, the optimisation of squeeze operational parameters should thus be performed based on the individual well or similar well and formation conditions.

8 SUMMARY AND CONCLUSIONS

In this paper, we have presented an extensive study of phosphonate scale inhibitor adsorption in both crushed and consolidated outcrop and reservoir rock materials. Static and dynamic adsorption

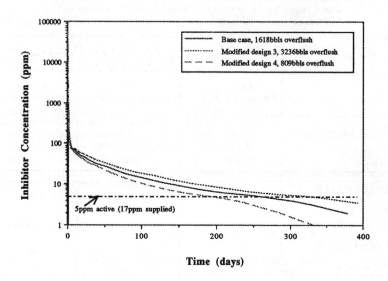

Figure 15   Comparison of squeeze inhibitor returns simulated with the
Original Design 1 and the modified designs altering overflush volume

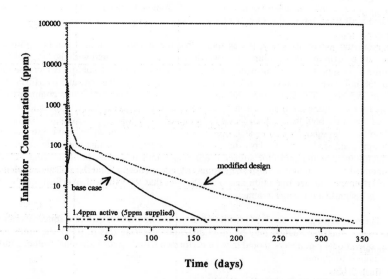

Figure 16   Comparison of squeeze inhibitor returns simulated with
the Original Design 2 and the modified design.

Table 3　　　**INHIBITOR SQUEEZE LIFETIMES PREDICTED FOR SQUEEZE TREATEMENT *1* IN FIELD *A* USING FLOOD F*3* DERIVED ADSORPTION ISOTHERM**

Treatment	Squeeze Lifetimes (days) at Threshold Inhibitor Concentrations (as active content)		
	50ppm (172ppm as supplied)	10ppm (34ppm as supplied)	5ppm (17ppm as supplied)
Original design: 20% inhibitor 1618bbls O/F	18	137	258
Modified design 1: 15% Inhibitor 1618bbls O/F	15	119	222
Modified design 2: 10% Inhibitor 1618bbls O/F	12	87	165
Modified design 3: 20% Inhibitor 3236bbls O/F (twice the base case)	25	171	320
Modified design 4: 20% Inhibitor 809bbls O/F (half the base case)	14	103	192

Table 4　　　**INHIBITOR SQUEEZE LIFETIMES PREDICTED FOR SQUEEZE TREATEMENT *2* IN FIELD *B* USING FLOOD F8 DERIVED ADSORPTION ISOTHERM**

Treatment	Squeeze Lifetimes (days) at Threshold Inhibitor Concentrations (as active content)		
	10ppm (34ppm as supplied)	5ppm (17ppm as supplied)	1.4ppm (5ppm as supplied)
Original design: Preflush 50bbls Main treatment 67bbls Overflush 790bbls	81	105	163
Modified design: Preflush 50bbls Main treatment 268bbls Overflush 395bbls	155	200	336

isotherms for the phosphonate/rock substrate were measured from the bulk solution tests and derived from the core flood inhibitor effluent data, respectively. The central importance of the dynamic adsorption isotherm, $\Gamma(C)$, was shown since this governs the form of the inhibitor return curve both in core floods and in field squeeze treatments. Bulk adsorption tests are of more use for establishing the sensitivities of inhibitor adsorption to various factors such as pH, $[Ca^{2+}]$, temperature, mineralogy etc. Dynamic isotherms were derived for field and outcrop cores and the application of these results to the design of an optimised "Field Squeeze Strategy" was illustrated using two field examples.

The specific conclusions from this study are as follows:

1) It is shown that static adsorption tests can be time-efficient and very useful to study broad adsorption *sensitivities* to various factors such as pH, $[Ca^{2+}]$, temperature and clay minerals. However, dynamic adsorption isotherms are more appropriate for describing interactions between inhibitor molecules and rock substrates under flowing conditions and they provides the vital information on adsorption in the very low threshold concentration region.

2) The dynamic isotherms derived from reservoir condition core flood data all show the characteristics of a steeply rising adsorption at concentration lower than ~ 100 ppm and a far more gradually increasing adsorption at C > ~ 200 ppm. These features are considered to be favourable to a long adsorption squeeze return.

3) The dynamic isotherms all show deviations from the exact mathematical forms of either the Freundlich or the Langmuir isotherms.

4) From an analysis of the shape of isotherms and the core flood inhibitor concentration return profiles, it is again clearly demonstrated that the inhibitor return curve is primarily governed by the shape of its adsorption isotherm, especially in the threshold concentration region.

5) The presence of clay minerals in the reservoir core material enhances inhibitor adsorption and increases the steepness of the isotherms at medium and high inhibitor concentration ranges. This could be significant to a squeeze lifetime if the inhibition threshold concentration required is rather high. On the other hand, this is less important if the inhibition level required for scale prevention is low (say, < 5 ppm).

6) Two case studies predicting and optimising scale inhibitor field squeeze treatments using experimental dynamic adsorption isotherm data were presented. These illustrate our modelling approach to the development of the "Field Squeeze Strategy" which may be summarised as follows:
    i)   carrying out the appropriate inhibitor core floods;
    ii) deriving dynamic isotherms from core flood data; and
    iii) using the isotherm data in the software to evaluate and optimise scale inhibitor performance in field squeeze applications.

7) Although the approach to the design of the "Field Squeeze Strategy" is the same, it has been demonstrated from the two modelled field examples that the sensitivities to operational parameters may be different in different cases. In one case, the suggestion was to increase the overflush size, in another a larger inhibitor slug and smaller overflush was suggested by the modelling.

ACKNOWLEDGEMENTS

The authors would like to thank the following companies for funding the current work of the Heriot-Watt University Oilfield Scale Research Group: Agip, Baker Performance Chemicals, BP

Exploration, Chevron, Elf, Exxon Chemicals, Kerr-McGee, Marathon, Shell, Statoil, Texaco and Total Oil Marine.

REFERENCES

1. P.J. Breen, H.H. Downs and B.N. Diel, <u>Royal Society of Chemistry Publication - Chemicals in Oil Industry: Developments and Applications</u>, Edited by P.H. Ogden, 1991.

2. G.M. Graham, K.S. Sorbie and I. Littlehales, Paper presented at the Water Management Offshore Conference held in Aberdeen, Scotland, October 6-7, 1993.

3. S.A. Hong and P.J. Shuler, <u>SPE Production Engineering</u>, November 1988, 597-607.

4. M.M. Jordan, K.S. Sorbie, P. Jiang, M.D. Yuan, L. Thiery and A.C. Todd (1994a), Paper presented at the NACE Annual Conference and Corrosion Show held in Baltimore, Maryland, February 28 - March 4, 1994.

5. M.M. Jordan, K.S. Sorbie, P. Jiang, M.D. Yuan, L. Thiery and A.C. Todd (1994b), Paper SPE 27389 presented at the SPE Formation Damage Control Symposium held in Lafayette, LA., February 7-10, 1994.

6. M.M. Jordan, K.S. Sorbie, P. Jiang, M.D. Yuan, A.C. Todd, K. Taylor, K. Hourston and K. Ramstad (1994c), Paper SPE 27607 presented at the SPE European Production and Operations held in Aberdeen, UK., March 15-17 , 1994.

7. A. Kan, P.B. Yan, J.E. Oddo and M.B. Tomson, Paper SPE 21714 presented at the SPE Production Operations Symposium held in Oklahoma, April 7-9, 1991.

8. A.T. Kan, X. Cao, X. Yan, J.E. Oddo and M.B. Tomson, Paper No. 33 presented at the NACE Annual Conference and Corrosion Show, Nashville, Tennessee, April 27 - May 1, 1992.

9. G.E. King and S.L. Warden, Paper SPE 18485 presented at the SPE International Symposium on Oilfield Chemistry held in Houston, TX., February 8-10, 1989.

10. K.O. Meyers, H.L. Skillman and G.D. Herring, <u>J. Pet. Tech.</u>, June 1985, 1019-1034.

11. P.J. Shuler, Paper SPE 25162 presented at the SPE International Symposium on Oilfield Chemistry held in New Orleans, March 2-5, 1993.

12. K.S. Sorbie, Paper presented at the Water Management Offshore Conference held in Aberdeen, Scotland, October 22-23, 1991.

13. K.S. Sorbie, A.C. Todd, R.M.S. Wat and T. McClosky (1991a), <u>Royal Society of Chemistry Publication - Chemicals in Oil Industry: Developments and Applications</u>, 199-214, Edited by P.H. Ogden, 1991.

14. K.S. Sorbie, M.D. Yuan, A.C. Todd and R.M.S. Wat (1991b), Paper SPE 21024 presented at the SPE International Symposium on Oilfield Chemistry held in Anaheim, CA, February 20-22, 1991.

15. K.S. Sorbie, R. M. S. Wat and A. C. Todd, <u>SPE Production Engineering</u>, August 1992, 307-312.

16.   K.S. Sorbie, M.D. Yuan, P. Chen, A.C. Todd, and R.M.S. Wat (1993a), Paper SPE 25165 presented at the SPE International Symposium on Oilfield Chemistry held in New Orleans, March 2-5, 1993.

17.   K.S. Sorbie, P. Jiang, M.D.Yuan, P. Chen, M.M. Jordan, and A.C. Todd (1993b), Paper SPE 26605 presented at the SPE 68th Annual Technical Conference and Exhibition held in Houston, TX., October 3-6, 1993.

18.   K.S. Sorbie and M.D.Yuan, *SQUEEZE IV: A Program to Model Inhibitor Squeeze Treatments in Radial and Linear Systems; User's Manual*, Department of Petroleum Engineering, Heriot-Watt University, May 1993.

19.   O.J. Vetter, J. Pet. Tech., March 1973, 339-353.

20.   M.D. Yuan, K.S. Sorbie, A.C. Todd, L.M. Atkinson, H. Riley and S. Gurden, Paper SPE 25165 presented at the SPE International Symposium on Oilfield Chemistry held in New Orleans, LA, March 2-5, 1993.

A New Assay for Polymeric Phosphinocarboxylate Scale Inhibitors at the 5 ppm Level

C. T. Bedford, P. Burns, and A. Fallah

SCHOOL OF BIOLOGICAL SCIENCES, UNIVERSITY OF WESTMINSTER, 115 NEW CAVENDISH STREET, LONDON, W1M 8JS, UK

W. J. Barbour and P. J. Garnham

PRODUCTION CHEMISTRY DEPARTMENT, SHELL UK EXPLORATION AND PRODUCTION, 1 ALTENS FARM ROAD, NIGG, ABERDEEN, AB9 2HY, UK

ABSTRACT: A new method of analysis of polymeric phosphinocarboxylates in well water returns has been developed, which involves the use of Solid Phase Extraction techniques and quantitative Size Exclusion High-Performance Liquid Chromatography. The method is capable of being used routinely to determine concentrations in produced waters at the 5 ppm level.

INTRODUCTION

Currently wells are squeezed with phosphinocarboxylate scale inhibitors (SI) every 180 days or so to ensure that protection is maintained. It is not necessary to re-treat until the phosphinocarboxylate content in the returned water has fallen to a level of about 5 ppm. It would be preferable to re-treat on the basis of SI concentration rather than on a calendar basis, but so far this has not been possible. The reason for this, generally, is the non-availability of trace analysis methods for phosphinocarboxylates that are capable of determining low ppm concentrations in produced waters. Clearly, if a reliable and sensitive method were available for phosphinocarboxylate determination in well water returns, the 180 day period could be extended, with resultant significant cost savings. Currently, assays of phosphinocarboxylates in produced waters can be made by procedures based on an adsorption-colorimetric method, but these procedures tend to be labour-intensive, susceptible to interferences and are of low selectivity.

This paper describes the development of a new method, using Solid Phase Extraction (SPE) and High-Performance Liquid Chromatography (HPLC), that can detect polymeric phosphinocarboxylates routinely at a concentration of about 5 ppm in well water returns. In addition, using a commercially-available vacuum manifold, the somewhat time-consuming clean-up by SPE has been semi-automated so that up to twelve samples can be processed simultaneously.

BACKGROUND

The polymeric phosphinocarboxylates that are used as scale inhibitors are composed of two medium-length polyacrylate chains interlinked by a phosphorus (V) atom in the form of a phosphinate. The general structure of the parent polyacrylates and of the derived phosphinocarboxylates are shown in **Scheme 1**.

$$-[CH_2\text{-}CH(CO_2Na)]_n -$$

POLYACRYLATES

$$
\begin{array}{c}
O \\
\parallel \\
-[CH_2\text{-}CH(CO_2Na)]_p - P - [CH_2\text{-}CH(CO_2Na)]_q - \\
\mid \\
ONa
\end{array}
$$

POLYMERIC PHOSPHINOCARBOXYLATES

Scheme 1 Structures of Scale Inhibitors.

The standard method for separation of polymers is Size Exclusion Chromatography (SEC), alternatively known as Gel Filtration. The basis of this method is selective adsorption within the pores of a cross-linked resin of the smaller rather than the larger polymer molecules. This means that in SEC the weakly adsorbed high-M_r compounds are eluted first. In the quantitative form of SEC, Size Exclusion High-Performance Liquid Chromatography (SEHPLC), the chromatogram obtained from a polymer solution usually consists of a broad peak, reflecting the fact that many 'polymers' contain a large range of M_r values.

For the water-soluble polyacrylates, an assay method employing SEHPLC with spectrophotometric (UV) detection was developed in 1988-89 at Shell Research's Thornton Research Centre.[1] The method depends for its success on the spectrophotometric detection at $\lambda=200$ nm of the weakly chromophoric carboxylate groupings. Under these conditions - at the limit of the UV range - the choice of organic solvent and the buffer that can be used as the mobile phase is severely limited. However, using 'HPLC Grade' solvents, methanol-water mixtures and acetonitrile-water mixtures, with or without phosphate buffers, have proved entirely successful.

At Thornton Research Centre (TRC), it was shown that this methodology was applicable to the assay of the polymeric phosphinocarboxylates. Moreover, mindful of the trace quantities of polymers that were to be assayed, early consideration was given to the inclusion of a concentration step in the protocol. The use of Solid Phase Extraction (SPE) was attractive, since it offered a means of removing the polymer from a large volume of produced water by sorption to a cartridge of 'reverse-phase' (C-18) adsorbent, and then recovering it by de-sorption in a much smaller volume of, say, (1:1) acetonitrile-water.

Of the several types of cartridge available commercially, the Waters SepPak range was chosen by TRC and found by them and by us to be of good all-round reliability. A typical Waters SepPak SPE cartridge, which has a Luer fitting at each end, is shown in **Figure 1**. Syringes, of varying capacity, containing the sample to be applied may be attached to the top of the cartridge, and an extra cartridge may, if necessary, be attached to the bottom of the cartridge.

In the final version of the assay method for polymeric phosphino-carboxylates developed by McKerrell and his colleagues at TRC, a sample of 25 ml of produced water containing small amounts (ppm) of polymeric phosphino-carboxylate scale inhibitor (e.g Servo UCA 371) which had been acidified to pH 2.5, was - via syringe - passed down a 'Classic' (0.36 g) SepPak C-18 cartridge; following an acid wash of the cartridge (which removed all of the water-soluble salts), the polymer was de-sorbed with 2.5 ml of (1:1) acetonitrile-water (via syringe) and assayed by direct injection of an aliquot (150 µl) into a SEHPLC System.[2] This achieves a tenfold concentration of the polymer, and at the same time - a welcome bonus - frees it of any inorganic contaminants (i.e effects a clean-up). The method was validated by spiking laboratory sea water with 5 - 50 ppm SI (Servo UCA 371) and obtaining >95% recoveries.[2]

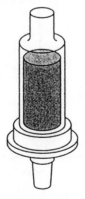

Figure 1 A Cartridge used in Solid Phase Extraction (SPE).

However, application of the method in the Shell Expro Production Chemistry Department, Aberdeen to field samples of produced waters spiked with 5 to 10 ppm SI ran into trouble because of 'interfering substances' in the final chromatogram. The problem was that these substances and the scale inhibitor had similar retention times on the SEHPLC column used, and often the size of the interfering peaks was more than tenfold that of the scale inhibitor. It was at this stage that work began on the problem at the University of Westminster.

SCALE OF THE PROBLEM

For the initial development work a sample of SI-free produced water from the Eider Field was used. We were thus able to spike aliquots of this sample with

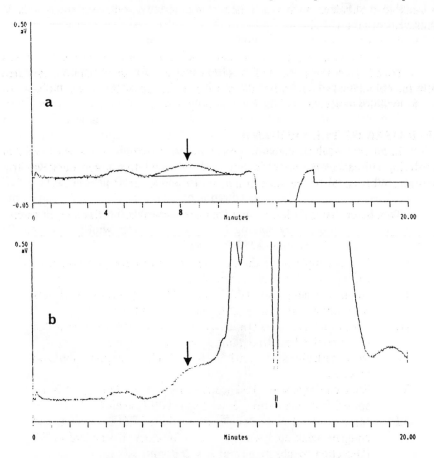

Figure 2 Chromatograms of **(a)** 50 ppm Servo UCA 371 in water **(b)** 'the concentrate' of an Eider produced water sample containing 5 ppm Servo UCA 371 after analysis by the TRC method (which includes a tenfold 'concentration' step).

known amounts of SI, usually 5 ppm, and then focus our attention on the interfering substances.

The scale of the 'interference' problem that needed to be overcome is best seen by inspection of the chromatogram of a typical spiked Eider sample. In **Figure 2 (a)**, the SEHPLC trace of a reference standard of 50 ppm Servo UCA 371 in water appears as a simple broad peak with a retention time of about 8 minutes. In **Figure 2 (b)**, the SEHPLC trace of the 'concentrate' of an Eider sample spiked with 5 ppm Servo UCA 371 after being subjected to the TRC method (which includes a tenfold concentration step) shows large interfering peaks at retention times similar to, and slightly longer than, that of the SI. The effect is to convert the SI peak to a shoulder on a much larger 'interfering' peak, and make reliable measurement an impossibility.

Nothing was known about the nature of these interfering peaks. What was clear at once from our inspection of the chromatogram was that because they eluted from the column at the tail-end of the broad polymer peak, they were probably of low to medium molecular weight (i.e. not polymeric).

APPRAISAL OF THE PROBLEM

In order to devise a clean-up procedure, we needed to know what types of interfering substances we were dealing with. We therefore asked ourselves this question. What could be the general nature of the interfering substances?

To answer that question we can tabulate some possibilities and, in each

---

I	Pass a sample of produced waters (30 ml) through a Whatman filter paper. (To remove solids).
II	Adjust the sample to pH 2.5 using 6M HCl (3 drops) (Converts SI to unionized form: $RCO_2^- \ Na^+ \Rightarrow RCO_2H$).
III	Apply the acidified sample (25 ml) to a Sep-Pak C-18 cartridge (SI is retained and inorganic salts pass through).
IV	Wash cartridge with 0.001 M HCL (3 ml). (Completes removal of salts).
V	Elute cartridge with (1:1) acetonitrile-water (0.25 ml). (SI is desorbed from the cartridge, ready for HPLC analysis).
VI	Inject 0.15 ml of the eluate on to the SEHPLC column, and compare peak area with peak area of external standard of SI. (Detection by absorbance at $\lambda = 200$ nm).

Scheme 2. A Summary of the TRC Procedure for Determination of Phosphinocarboxylate Scale Inhibitors in Well Water Returns

case, note whether they are, or are not, absorbed (together with the SI) on the C-18 cartridge at pH = 2.5. It will be recalled that the TRC method (summarised in **Scheme 2**) calls for initial acidification of the produced water sample to pH 2.5. This procedure, as can be seen from **Step II** in **Scheme 2**, converts the polymer 'salt' into its free acid form - rendering it lipophilic and readily adsorbable on a C-18 reverse-phase cartridge.

In general terms, we can consider as candidate interfering substances an organic acid (say a carboxylic acid, RCOOH), an organic base (say an amine, R_3N), and a neutral compound (X). We need to record their state of ionization at pH = 2.5, for ionic compounds will not be absorbed by the lipophilic C-18 phase.

These data appear in **Table 1**, from which we can see that amines cannot be the interfering species, for they will be ionized at pH = 2.5 and pass right

Table 1: Fate of organic compounds on C-18 cartridge at pH = 2.5.

Compound	Absorption on C-18 Cartridge
RCOOH	YES
R_3NH^+	NO
X	YES

through the C-18 cartridge. However, neutral compounds and acidic compounds are possible candidates for the interfering species.

If we now inspect the data in **Table 2**, we can see that a method of removal of neutral interfering compounds is accessible by a simple manipulation of the pH

Table 2: Fate of organic compounds on C-18 cartridge at pH = 7.5.

Compound	Absorption on C-18 Cartridge
RCO_2Na	NO
R_3NH^+	NO*
X	YES

* Assuming that pK_a = 9.5

of the solution applied to the C-18 cartridge. At pH = 7.5, the SI will be poly-anionic and will not be retained. Neither will the carboxylic acid, for it will be present as the ionic carboxylate, RCO_2^- Na^+. But the neutral interfering substances, X, will be absorbed.

FORMULATION OF A SOLUTION TO THE PROBLEM

What is needed, therefore, to remove neutral interfering compounds is an extra clean-up step. And a very simple one. The produced water, as received, with its pH= 7.5 (approx) is passed through a C-18 cartridge - *prior* to adjustment of the pH to 2.5 and the absorption of SI on to another cartridge.

In the event, we found that we needed two cartridges (2 x 360 mg) to effect the clean-up of a 30 ml sample of Eider waters. We proved this by passing the sample down three cartridges sequentially and then, by de-sorption and SEHPLC-assay of each, showing that interferences were present in the first and second cartridges, but none in the third cartridge. The chromatograms that were obtained in this experiment are shown in **Figure 3**.

On the basis of these encouraging results, an extra clean-up step was inserted into the standard procedure between Step I and Step II, which we called Step IA. The procedures for Step IA - and of the appropriately revised Step II - are shown in **Scheme 3**.

IA.   Apply the sample of produced water (30 ml) sequentially to two Sep-Pak C-18 cartridges. (Neutral compounds are retained and salts and SI pass through). Discard the two cartridges.

II.   Adjust the eluate from Step IA to pH 2.5 using 6 M HCl (dropwise).

Scheme 3 Steps IA and II in Revised Procedure for SI Analysis.

TESTING THE REVISED PROCEDURE

The revised procedure was then applied to produced waters from the Eider Field spiked with 5 ppm SI. It gave very encouraging results, as can be seen from the reproduced chromatograms in **Figure 4**, which are of (a) a reference standard in de-ionized water of 50 ppm Servo UCA 371 and (b) the 'concentrate' of an Eider sample containing 5 ppm Servo UCA 371 which had been subjected to the revised Method (which includes a tenfold 'concentration' step). Clearly, there is

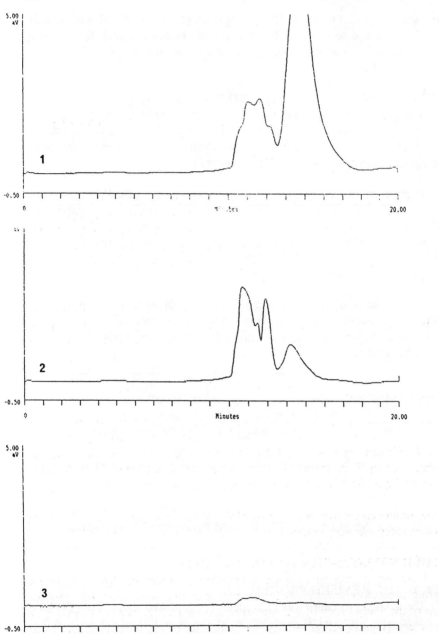

Figure 3 Chromatograms of the interfering substances retained by three C-18 Sep-Pak cartridges when Eider produced water (30 ml, pH=7.8) was passed sequentially through one **(1)**, two **(2)**, and three **(3)** of them (elution of each cartridge was effected with 2.5 ml (1:1) acetonitrile-water).

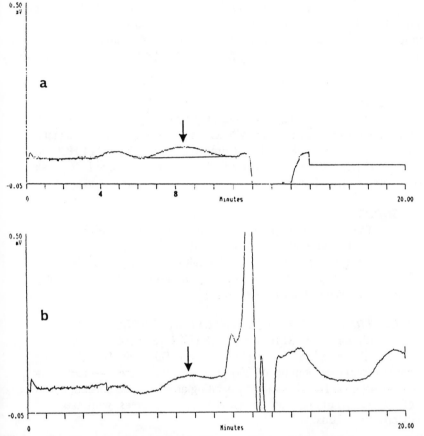

Figure 4 Chromatograms of **(a)** 50 ppm Servo UCA 371 in water **(b)** 'the concentrate' of an Eider produced water sample containing 5 ppm UCA 371 after analysis by the *revised* TRC method (includes a tenfold 'concentration' step).

a striking improvement over the original TRC Method (see **Figure 2**), and now the SI peak is accessible and measurable.

FURTHER REFINEMENTS TO THE PROCEDURE

However, as can be seen from the chromatogram in **Figure 4 (b)** the interfering substances have not been removed entirely. Since, from our previous analysis of the likely nature of the interfering substances, we had been able to rule out basic compounds such as amines, the remainder of the interfering substances must be acidic compounds. If the compounds were carboxylic acids, RCOOH, they would be difficult to remove because they will adsorb to, and de-sorb from, the C-18 cartridge under the conditions used to adsorb and de-sorb the SI.

However, we suspected that weakly acidic phenols might be the culprits. Accordingly, we made a slight adjustment to the protocol of the method by effecting the clean-up at pH 5.5 rather than at pH 7.5. This higher acidity had the effect of converting any weakly acidic phenols present in their 'salt' forms

---

1. CLEAN-UP

30 ml of a filtered sample of produced waters, having been transferred from a measuring cylinder into a beaker, is, with magnetic stirring, acidified to pH 5.5 by the dropwise addition of 1.0 M HCl and is then (via a syringe) passed through a pre-wetted Sep-Pak "Environmental" Cartridge (1 g).

2. SORPTION OF SI AND WASH

The eluate from (1) is, with magnetic stirring, acidified to pH 3.5 by the dropwise addition of 1.0 M HCl, and a pipetted aliquot (25 ml) is then (via a syringe) passed through a pre-wetted Sep-Pak C-18 "Classic" cartridge (0.36 g). The cartridge is washed with water (15 ml) and then (5:95) acetonitrile-water (5 ml).

3. ELUTION OF SI AND SIZE EXCLUSION HPLC ANALYSIS

Elution of the SI from the C-18 cartridge is effected with (1:3) acetonitrile-phosphate buffer (pH=7) (1 ml).[This is the HPLC moving phase.] Replicate aliquots of 250 µl of this eluate are injected into a Size Exclusion HPLC System* operating at 30 °C, and concentrations of SI are determined by comparison of peak areas with those of external standards.

---

* Size Exclusion HPLC System
Merck Column Thermostat, T-6300; Merck-Hitachi Pump, L-6200; Perkin Elmer Autosampler, ISS-100; Perkin Elmer Diode Array Detector, LC135; Jones Chromatography Data System, JCL 6000 Ver 4.23;
Column: 0.3 m x 7.8 mm TSK Gel G-2000 SW_{XL}, equipped with a guard column 0.04 m x 6.0 mm TSK Gel G-2000 SW_{XL};
Mobile Phase: 0.025 M KH_2PO_4 / 0.15 M KOH (pH 7.0) incorporating 25% acetonitrile (Rathburn 'S' Grade).
Flow Rate: 1 ml/min.

Scheme 4   Final Version of the SPE/SEHPLC Procedure for Analysis of Polymeric Phosphinocarboxylates.

into their uncharged forms (PhO⁻ Na⁺ ⟹ PhOH), and thus into a form readily adsorbed onto a C-18 cartridge. This procedure removed some more of the interfering substances, but some persisted. (We, of course, had checked that none of the SI was adsorbed to the C-18 cartridge at pH 5.5.)

From trial and error, we introduced several other improvements to the general method, which either reduced further the amounts of interfering substances, or, importantly, reduced the time of analysis - which had been increasing unacceptably. The final version of the Procedure, which includes details of the SEHPLC System, is shown in **Scheme 4**.

In the **Clean-up Procedure (Step 1, Scheme 4)**, two small cartridges were replaced with one of larger capacity (called an 'Environmental' cartridge because of its widespread use in the trace analysis of water-borne environmental contaminants). This reduced the time of analysis. For the **Sorption of SI (Step 2, Scheme 4)** the sample was adjusted to pH 3.5 prior to passage through a C-18 cartridge (having been found to be an improvement on adjustment to pH 2.5, since lesser amounts of interfering substances were encountered). **Washings** of the cartridge containing the sorbed SI with first water (15 ml) and then (5:95) acetonitrile-water (5 ml) were important in removing virtually all of the remaining interfering substances. **Elution of SI (Step 3, Scheme 4)** from the C-18 cartridge was effected with the SEHPLC moving phase, which had been found to reduce 'back-peaking' in the chromatograms. Also by reducing the amount of eluent from 2.5 ml to 1.0 ml, the concentration factor - and hence the sensitivity of the method - increased from 10-fold to 25-fold. Finally, injection of 250 μl, rather than 150 μl, into the SEHPLC system further increased the sensitivity of the method.

PUTTING THE FINAL VERSION OF THE NEW PROCEDURE TO THE ULTIMATE TEST

Legend had it that the field with produced waters containing the highest amounts of interfering substances was Cormorant. It was decided therefore to submit the new method to its sternest test - the analysis of SI in Cormorant well water returns. If it could cope with this sample, it would, it was considered, be robust enough to cope with any other from any location.

Thus two samples of Cormorant produced waters spiked with 0 ppm (a control) and 5 ppm Servo UCA 371 were analysed by the final version of the new method. Duplicate analyses were carried out for each sample. The results are shown in **Table 3**, with peak areas representing concentrations of SI in external standards and in test samples. Overall, the method effects a concentration of the sample by a factor of 25, and therefore 5 ppm in the original water sample becomes 125 ppm in the final aqueous 'concentrate'.

Table 3. Analysis of Cormorant Waters spiked with SI (Servo UCA 371).

	Peak Area	% Recovery	% Mean Recovery
Standard, 125 ppm	(a) 727		
	(b) 655 mean 694		
	(c) 700		
CW + 5 ppm SI	(a) 669	96%	
	(b) 612	88%	92%
CW + 0 ppm SI	(a) 39	6% (interfering	
	(b) 29	4% substances)	"5%"

Excellent results, with a mean recovery of 92%, were obtained for the spiked 5 ppm samples. As expected, the unspiked control samples showed no peak due to the presence of SI, but a general "background" that was observed was acceptably low and amounted to about 5%. The HPLC chromatograms which were obtained from these samples and those of reference standards of SI are shown in **Figures 5 and 6**.

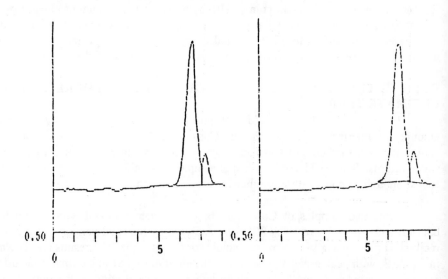

Figure 5 SEHPLC traces of two of three replicates of a 125 ppm standard of Servo UCA 371.

In **Figure 5**, the chromatograms of two of the three replicate samples of a 125 ppm Servo UCA 371 standard show good reproducibility. There is also a small peak at R_t = 7.2 min, which was (easily) programmmed not to be included in the integral (see manually-simulated base line and vertical 'drop' in each chromatogram of **Figure 5**).

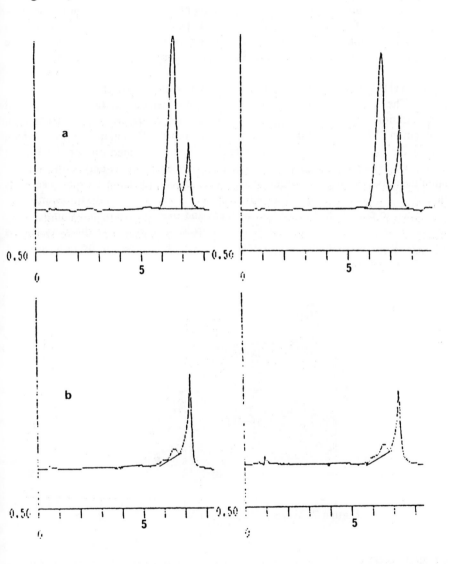

Figure 6   SEHPLC traces of Cormorant well waters: **(a)** spiked with 5 ppm Servo UCA 371 and **(b)** spiked with 0 ppm Servo UCA 371 (controls).

This 'extra' peak at R_t = 7.2 min also appears in the HPLC traces of the duplicate 5 ppm samples, as can be seen from **Figure 6 (a)**. Now the cleaned-up sample is nearly as 'clean' as the standard itself! (Compare **Figure 5** and **Figure 6 (a)**.)

This is further confirmed by the results obtained with the control samples. As can be seen from **Figure 6 (b)**, the 'extra' peak at R_t = 7.2 min is present in the HPLC chromatograms of the duplicate control samples, but, importantly, the "background" peaks at the R_t of the SI amount to very little (approx 5%).

SEMI-AUTOMATION OF THE CLEAN-UP PROCEDURE
The time-consuming part of the new method is the clean-up and concentration steps using the Solid Phase Extraction cartridges. This is a common problem in SPE work, and there already exist commercial vacuum manifolds which enable the process to be semi-automated. We have evaluated one of the Waters Manifolds, shown in **Figure 7**, which can process up to twenty four samples simultaneously. Using the large-capacity cartridges with large solvent reservoirs that were needed for the clean-up step, we found the manifold became too congested to cope with twenty-four of them and only twelve of the stations were useable in practice. However, overall the manifold worked very well and a protocol has now been devised, and is currently under test, for its use in the processing of twelve samples at a time.

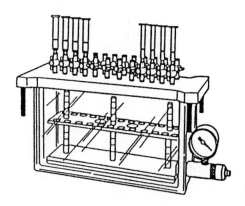

Figure 7 Waters 24-Position Vacuum Manifold.

CONCLUSION
In summary, the new method that we have developed has now been validated using typical 'worst' samples and has been shown to be capable of

determining phosphinocarboxylate SI in produced waters at the level of 5 ppm. With the benefit of the semi-automation provided by the vacuum manifold, the method is now ready to be evaluated against existing methods in a full field trial.

ACKNOWLEDGMENTS

We thank Peter Hanaway, University of Westminster, for expert technical assistance with the SEHPLC System.

REFERENCES

1. E.H. McKerrell, D. Worrall and A Lynes, *Development of an analytical method for the determination of polyacrylate scale inhibitors in well water returns*. Shell Research Ltd, Thornton Research Centre Group Report, TNGR.89.013.

2. E.H. McKerrell, *Development of an analytical method for the determination of polymeric phosphinocarboxylic acid scale inhibitors in well water returns*. Shell Research Ltd, Thornton Research Centre Group Report, TNGR.89.109.

Using Statistical Experimental Design to Optimise the Performance and Secondary Properties of Scale Inhibitors for Downhole Application

G. E. Jackson, G. Salters, and P. R. Stead

PETROLITE LIMITED, LIVERPOOL, UK

B. Dahwan and J. Przybylinski

PETROLITE CORPORATION, ST. LOUIS, MISSOURI, USA

1. INTRODUCTION

Long term control of inorganic scale deposition downhole in high volume wells can be achieved using "squeeze" treatment of "threshold" inhibitors. The development of appropriate inhibitors for downhole application and the design of a successful application involves the consideration of a number of factors [1].

The most widely used chemicals for downhole treatment are aminomethylene phosphonic acids, and carboxylic acid containing polymeric materials of about 3,000 to 20,000 molecular weight.

The phosphonic acid inhibitors may be made from many different amines. Various fractions of the amine hydrogens may be substituted with methylene phosphonic acid groups. The degree of substitution affects their performance as inhibitors and their secondary properties such as adsorption-desorption characteristics, and brine solubility. Generally the higher the degree of substitution, the more effective the inhibitor, but the less tolerant the inhibitor becomes to high levels of divalent cations. The brine solubility is extremely important in the design of squeeze programmes.

Similar trends are seen with polymeric inhibitors as the molecular weight of the polymer increases, the brine tolerance decreases. Increasing the number of functional groups per unit chain length may improve its inhibition performance but may make it more insoluble in brines.

Scale Inhibition

The mechanism of mineral scale deposition is generally believed to involve adsorption of inhibitors and poisoning of active growth sites on the crystal surface. Anionic groups, such as phosphonates and carboxylates, adsorb at cationic sites and inhibit the growth of mineral scale crystals even when present at sub-stoichiometric concentrations[2,3].

Inhibitors may also adsorb on the crystal surface and cause a change in the morphology of the growing crystal. It has been shown[4] that polymeric inhibitors such as polyacrylic acids are incorporated into the growing crystal and this leads to considerable distortion of the crystal lattice.

Inhibitor Effectiveness

The pH of the brine system downhole has a major effect on the efficiencies of scale inhibitors. The pH affects the degree of deprotonation of the active functional group which governs the extent to which the inhibitor will adsorb on the growing mineral scale crystal. It also affects the adsorption and desorption on the formation, and the solubility of the inhibitor in the brine.

It has been claimed that polymeric inhibitors perform better than phosphonates against sulphate scales, and barium sulphate in particular, in conditions of low pH and high barium to sulphate ratios. Polymers are also claimed to be more thermally stable under downhole conditions than the more conventional inhibitors.

However, there are a number of snags associated with the use of polymers for squeeze treatments. The major drawback is the inability to monitor low levels of inhibitor in the produced fluids. Also kinetic desorption studies [5] indicate that polyacrylates do not desorb at a rate great enough to give an acceptable residual level of inhibitor. The marginal brine solubility of polymers can lead to problems in placing the inhibitor in the formation and may lead to plugging of the formation.

2.   BACKGROUND

Initial Development

Some years ago a project was set up to synthesise phosphorus containing polymers which would be effective at inhibiting barite under low pH conditions for downhole application. The polymers developed were N-phosphonomethylated amino-2-hydroxy propylene polymers similar to those revealed in U.S. Patent 4857205.

Various amines were investigated as components of the polymer backbone structure, and the one that gave the best performing products identified. The manufacturing process variables were then varied to give the product with the best scale inhibition properties. The particular property we optimised was the inhibition of barite formation under low pH conditions. These conditions are found in a number of North Sea fields, and in fields which are being flooded with CO_2.

This optimised structure was also found to inhibit calcite, gypsum, and barite, as well as, or better than commercially available scale inhibitors at more neutral pH's. Unfortunately, the polymer, like many others, had poor brine solubility, which could lead to problems if it was attempted to squeeze it into a hot formation, with the potential for plugging the well and causing formation damage.

Clearly the polymer had to be modified further to achieve superior low pH barite scale inhibition activity, very good neutral pH scale inhibition activity, and brine solubility as good or better than conventional phosphonates, which have been squeezed successfully for many years in the North Sea without any problems.

3.   EXPERIMENTAL DESIGN

Introduction

Early preliminary work had identified a polymer backbone structure which gave products which were much more soluble than the backbone we initially decided to work with, but did not perform as well. It was decided

to use experimental design to increase the performance
of this more soluble product without sacrificing its
excellent brine tolerance. A statistical experimental
design programme was used to help plan and analyse the
experiments.

Variables

In the manufacture of the polymer back bone, an
amine, epichlorhydrin and other components are reacted
in various proportions to arrive at the final product.
From a knowledge of the chemistry involved, and
experience, it was felt that some of these should
remain fixed, and that it is best if others are varied
in proportion to some others. This left us with only
two truly independent components, epichlorhydrin, A,
and a solvent, B.

Increasing the ratio of epichlorhydrin "A"
increases the molecular weight, and the viscosity of
the product.

Increasing the solvent "B" decreases viscosity and
may affect the branching of the polymer. Not all
combinations of A and B were possible. Very high
levels of A could not be used with very low levels of B
otherwise the polymer became too viscous to process.
With the aid of a statistical design programme, a set
of experiments were identified. This set consisted of
eight experiments consisting of seven different
combinations of A and B with one repeat.

4.   RESULTS

Experimental

The eight products which were prepared, are listed
in Table 1. Compound numbers are given for easy
reference later. They were prepared and their pH 4
barite inhibition activities were determined using a
seeded crystal precipitation test [6]. The brine
solubilities were also tested using a synthetic brine
system which simulates an 80/20 mixture of a particular
North Sea formation water and sea water [6]. The
results are shown in Table 1.

TABLE 1
COMPOSITIONS AND TEST RESULTS

COMPOUND NUMBER	A g mol^{-1}	B g mol^{-1}	BARIUM RETAINED mgl^{-1}	SOLUBILITY (%)
92	18.5	41.2	12.7	93.2
93	18.5	20.6	16.7	94.7
94	27.8	41.2	16.9	81.0
95	27.8	20.6	23.9	61.7
96	37.0	41.2	20.6	88.7
97	37.0	20.7	25.9	72.4
98	18.5	0.0	26.2	84.5
99	18.5	0.0	25.1	93.0

TABLE 2
COMPONENTS ANALYSIS OF VARIANCE FOR BARIUM RETAINED

SOURCE	DEGREES OF FREEDOM	SUM OF SQUARES	MEAN SQUARE	F-RATIO	LACK OF SIGNIFICANCE
CONSTANT	1	3538.00			
A	1	72.94	72.94	47.5	0.001
B	1	161.90	161.90	105.3	0.000
RESIDUAL	5	7.69	1.54		

R^2 = 0.9574

TABLE 3
COMPONENTS ANALYSIS OF VARIANCE FOR PERCENT SOLUBILITY

SOURCE	DEGREES OF FREEDOM	SUM OF SQUARES	MEAN SQUARE	F-RATIO	LACK OF SIGNIFICANCE
CONSTANT	1	1252.6773			
A	1	501.3293	501.3292	14.33	0.0193
A*B	1	243.4807	243.4807	6.96	0.0577
A**2	1	404.1777	404.1777	11.56	0.0273
RESIDUAL	4	139.8993	34.9748		

R^2 = 0.8515

Analysis

The data was analysed statistically to minimise the effect of scatter or "noise" and gave the best interpretation and also an estimate of the confidence level of the conclusion.

The amount of barium remaining in solution after the inhibition test and the solubility of the product are the responses in the statistical analysis. The responses are controlled by the amounts of components A and B, which are the predictors. Initially both responses were considered to be second degree functions of the predictors. When the data were analysed, only the linear terms in A and B were found to be statistically significant predictors of barite inhibition. The solubility response was fit best by a function consisting of first and second degree terms in A and the cross term, AB.

The summary components analyses of variances for the two responses are shown in Tables 2 and 3. The barium function fits the barite inhibition response very well; we are confident that both A and B are significant predictors of scale inhibitor performance. The solubility response does not fit the solubility function as well. There is a 1 in 17 chance that the predictor AB is not significant.

Performance Predictions

The functions fit to the response can be used to generate a contour diagram. The diagram is shown in Figure 1. Barite inhibition varies linearly. Best performance is predicted to come from products with no B and 37 units of A. Unfortunately products in this region can not be made because the polyamine substrate becomes too viscous to process. With no B the maximum allowable amount of A is about 28 units. With 37 units of A, at least 14 units of B must be used. Any compositions along a diagonal line should give the same performance as any other composition on that line. Table 4 gives the predicted amount of barium retained in solution for compositions roughly paralleling one of the diagonal lines. The 95% confidence interval is also given.

Solubility Predictions

The solubility behaviour is more complex than the performance behaviour. The most soluble products seem

FIGURE 1

Predicted Values

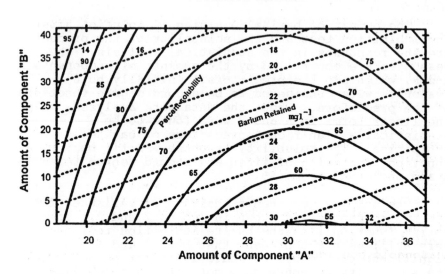

FIGURE 2

Predicted Values

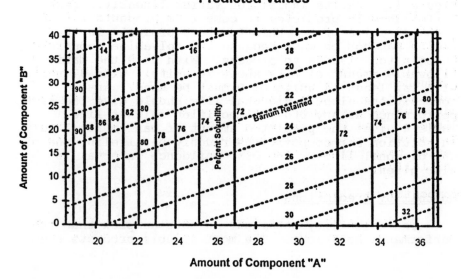

to be made with 17 units of A and 41 units of B. The influence of B on solubility, however, must be interpreted with caution. This is illustrated in Table 5, which gives predicted solubilities for the same compositions shown in Table 4. The 95% confidence interval is so large that it can not be said with any reasonable degree of certainty that any composition is more or less soluble than another in this range.

Since B was not found to be individually significant, the cross term AB was dropped. When this is done solubility is considered to be only a function of A and A^2.

Considering only A and A^2 generates the simplified prediction shown in Figure 2. Maximum solubility is predicted at 17 units of A, and minimum solubility near 30 parts A. An analysis of the variance (Table 6) shows that the fit is now worse than when the term in AB is included. But an analysis of confidence levels for the predicted shows that we can be more confident of the predicted solubilities, although the 95% confidence intervals is still too large for us to have reasonable confidence in the predictions.

Given the poor statistics of the solubility fits, it is possible that neither A nor B influences solubility. The data in Table 6 shows that there is at least a 1 in 12 chance that this is true. If this is the case, the predicted solubility is 83.6% with a 95% probability that it lies in the range of 75.6 to 91.7%.

5.   INHIBITOR DEVELOPMENT

Checking Predictions

Since it was not clear that A has any significant influence on solubility, the most desirable composition was initially chosen on the basis of barite inhibitor performance. We could choose any composition at the diagonal line from 28 units of A, zero B to 37 units of A, 14 units of B. These compounds should perform better than any of the eight compounds prepared initially. Compound 106 was prepared with A = 27.8, B = 0.0. It was found to outperform all the other compositions. This prediction was confirmed.

To double check the predictions given in Tables 4

TABLE 4
PREDICTIONS AND 95% CONFIDENCE INTERVALS FOR BARIUM RETAINED

AMOUNT OF A	95% INTERVAL	AMOUNT OF B		
		0.0	7.0	14.0
A = 27.8	LOWER	25.6	24.2	22.6
	PREDICTED	29.2	27.1	24.9
	UPPER	32.8	30.0	27.1
A = 32.4	LOWER		25.4	23.9
	PREDICTED		29.1	26.9
	UPPER		32.1	29.9
A = 37.0	LOWER			24.9
	PREDICTED			28.9
	UPPER			32.9

TABLE 5
PREDICTIONS AND 95% CONFIDENCE INTERVALS FOR PERCENT SOLUBILITY

AMOUNT OF A	95% INTERVAL	AMOUNT OF B		
		0.0	7.0	14.0
A = 27.8	LOWER	21.8	29.9	37.3
	PREDICTED	56.8	260.1	63.4
	UPPER	91.7	90.2	89.4
A = 32.4	LOWER		26.9	36.5
	PREDICTED		58.6	62.4
	UPPER		90.3	88.4
A = 37.0	LOWER			40.6
	PREDICTED			69.9
	UPPER			99.2

TABLE 6
COMPONENTS ANALYSIS OF VARIANCE FOR PERCENT SOLUBILITY - REVISED

SOURCE	DEGREES OF FREEDOM	SUM OF SQUARES	MEAN SQUARE	F-RATIO	LACK OF SIGNIFICANCE
CONSTANT	1	1051.3257			
A	1	356.0526	356.0526	4.64	0.0837
A**2	1	308.5087	308.5067	4.02	0.1012
RESIDUAL	5	383.3800	76.6760		

R^2 = 0.5932

TABLE 7
PREDICTIONS AND 95% CONFIDENCE INTERVALS FOR
PERCENT SOLUBILITY - REVISED

95% INTERVAL	AMOUNT OF A		
	27.8	32.4	37.0
LOWER	46	51	55
PREDICTED	71	72	80
UPPER	96	93	105

and 5, three additional inhibitors were prepared corresponding to the compositions A=37.0, B=14.0, A=32.4, B=7.0, and A=27.8, B=0.0. These are, respectively, compounds 162, 163 and 164. They were predicted to have nearly identical performance. Their solubilities should have been slightly different if A or B were a significant predictor of solubility. For comparison two compounds prepared earlier were tested along with these three. Compound 106 should have been identical to 164 and compound 99 should have been a poorer performer.

The compounds were tested using the pH seeded barite test [6]. Results are shown in Figure 3. Although compounds 162, 163, 164 and 106 did not yield identical test results, they were very similar. The differences are judged to be within experimental error. Compound 99 was definitely inferior to the others. The performance predictions were substantiated.

Solubilities

Compounds 162, 163, 164, 99 and 106 were formulated into finished products. These were labelled respectively 1A, 1B, 1C, 1D and 1E. Their solubilities were compared with that of a phosphonic acid inhibitor, which has been used successfully for many years in the North Sea. The solubilities were determined using a method that simulates the preparation of a squeeze solution [6]. Figure 4 shows that the solubilities of these compounds are equal to a commonly used phosphonic acid, SP-X.

6.   EXPERIMENTAL SCALE PREVENTATIVE XSP-1182

Solubility

We have since done many other solubility tests in oilfield brines. In virtually every case the polymer chosen for development was found to have solubility as good as or better than a conventional phosphonic acid which has been successfully squeezed in the North Sea for many years. Consequently the data indicates that this polymer will not cause problems from a brine solubility/formation damage standpoint.

FIGURE 3   **Inhibitor Effectiveness in a pH 4, 40C Barite Test**

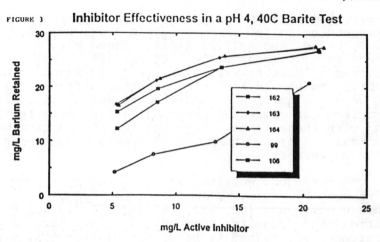

FIGURE 4   **Solubility** of Inhibitors in a Simulated Squeeze at **120C**

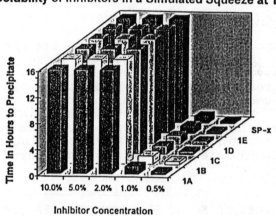

FIGURE 5   **Performance Testwork in Calcium Brine Systems**

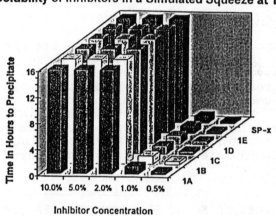

Performance Testing

The efficiency of an inhibitor depends on the type of scale to be inhibited, the brine composition, and the temperature. The rankings of inhibitors can often be changed by adjustments of these three factors. In particular it is not unusual for polymers to work better at low temperatures and phosphonates at high temperatures[7]. For proper selection of inhibitors it is important to match test brines and physical conditions to field conditions.

The development product XSP-1182, and a commercial phosphonic acid type inhibitors, SP-X, and a substituted polyacrylate SP-Y were tested for calcite, gypsum, and barite scale inhibition at 95°C in bottle tests [6]. Results are shown in Figures 5 and 6. XSP-1182 outperformed conventional products in all these tests. Also when soluble iron was incorporated in the barite precipitation tests, the development product XSP-1182 showed good tolerance to iron and gave improved performance over the commercially available products.

Adsorption Characteristics

Selection of an appropriate inhibitor depends on more than just performance and solubility. If an inhibitor is to be squeezed, one of the most important properties is how well it returns in the produced fluids. Although there is still controversy surrounding the exact mechanisms by which inhibitors are retained and later released into the produced fluids, there is general agreement in the more recent literature that adsorption plays an important, if not dominant role. The solubilities of calcium salts of inhibitors are generally too great to account for the levels of inhibitors returning in the produced fluids, and reservoir related mechanisms do not depend on the properties of inhibitors.

Almost all of the oil reservoirs in the world consist of either carbonate or siliceous type materials. Scale inhibitor squeezes are often found to perform better in carbonate reservoirs, so tests were done on the more difficult case of siliceous reservoirs. Studies of the adsorption of this new polymer on a siliceous material were carried out using a powdered silica gel because its large surface area decreased experimental error. Tests were run in glass beakers thermostatically controlled at 60°C for one

FIGURE 6

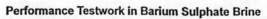

Performance Testwork in Barium Sulphate Brine

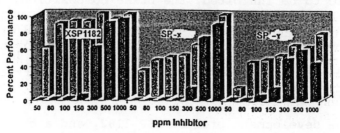

■ BaSO4 & Iron   ▨ BaSO4

FIGURE 7

Adsorption of XSP1182 on Silica Gel

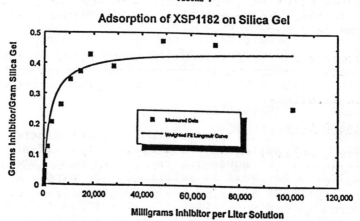

FIGURE 8

Adsorption of SP-x   on Silica Gel

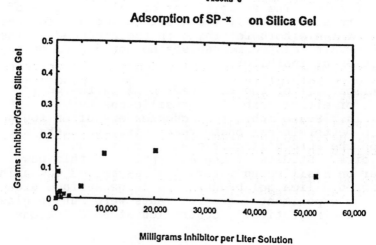

hour. The results are shown in Figure 7. A Langmuir Isotherm fitted the data much better than a Freundlich Isotherm.

Tests on the kinetics of the adsorption process showed that adsorption was complete in a minute or less. Extending the tests to sixteen hours made no difference. Actual field experience of squeezing shows that long shut-in times of twenty four hours or longer can be beneficial, and extend the squeeze life, which indicates that mechanisms other than adsorption on the formation are important.

A conventional commercial phosphonic acid was studied under identical conditions. It adsorbs more weakly, as shown in Figure 8. Because of the weaker adsorption there is a great deal more scatter in the data. Neither the Freundlich nor the Langmuir Isotherm fits the data very well.

Because XSP-1182 adsorbs at least three or four times more strongly than a conventional phosphonic acid, it is likely that XSP-1182 will give longer squeeze lifetimes.

Residual Determinations

The inability to easily determine returning concentrations of inhibitors after a squeeze treatment is one of the major drawbacks of most polymers. However, XSP-1182 contains a substantial amount of phosphorus in the molecule and residual levels can be accurately measured down to a detection limit of 1 ppm. The analytical method is the same as that used for conventional aminomethylene phosphonic acids.

7. CONCLUSIONS

The careful use of statistically designed experiments can be very helpful in the development of products which must simultaneously fulfil diverse criteria.

Statistical conclusions must be interpreted with caution, never losing sight of the statistical uncertainty in the answer.

We have developed a product, XSP-1182, with a superior set of properties for squeezing in the North

Sea. These include, but are not necessarily limited to:

a) Good inhibitory action against all types of scale, particularly barite in low pH conditions.
b) Good compatibility with oilfield waters.
c) Good adsorption properties.
d) Easy determination of returning residuals.

REFERENCES

1) G.E. Jackson : "Downhole Scale Prevention in High Volume Wells by Squeeze Treatment of Inhibitors" 2nd Symposium "Oilfield Chemicals", University of Clausthal, Germany, September 1984

2) Seung Tsuen Liu & D.W. Griffiths : "Adsorption of Amino Methylene Phosphonic Acids on the Calcium Sulfate Dihydrate Crystal Surface" SPE 7863, 1979, SPE of AIME, International Symposium "Oilfield Chemistry", Houston, Texas, January 1979

3) W.H. Leung & G.H. Nancollas : "Nitrilotri-methylenephosphonic Acid Adsorption on Barium Sulphate Crystals & its Influence on Crystal Growth" - J. Crystal Growth 44, 163 - 167, 1978

4) J.E. Crawford & B.R. Smith : Adsorption of Polyelectrolytes during Crystallisation of Inorganic Salts" J. Colloid Interface Science, (1966) 21, 623 - 625

5) J. Przybylinski : "Adsorption & Desorption Characteristics of Mineral Scale Inhibitors as Related to the Design of Squeeze Treatments" SPE 18486, SPE Symposium, Houston, Texas, February 1989

6) Detailed test procedures for all the experimental work are available on request from Petrolite Ltd., Kirkby, Liverpool.

7) M.C. Cushner, J.L. Przybylinski & J.W. Ruggeri : "How Temperature and pH Affect the Performance of Barium Sulfate Inhibitors" Paper 428, NACE Corrosion 88, 21-25 March 1988, St. Louis, Missouri, U.S.A.

8) D. Redmore, B. Dhawan, J.L. Przybylinski : "Method for Inhibition of Scale Formation" U.S. 4,857,205, 15 August 1989.

Oilfield Reservoir Souring – Model Building and Pitfalls

R. D. Eden

CAPCIS LTD, BAINBRIDGE HOUSE, GRANBY ROW, MANCHESTER, M1 2PW, UK

P. J. Laycock

DEPT. OF MATHEMATICS, UMIST, SACKVILLE STREET, MANCHESTER, M60 1QD, UK

G. Wilson

POLYMATHEMATICA, PO BOX 1066, BIRCHWOOD WARRINGTON, WA3 4PG, UK

1 BACKGROUND

Increasing daily yields of H_2S in produced reservoir fluids, conventionally known as souring, are now commonly attributed to biogenic activity in seawater flooded oil formations (1-5). Biological activity, by sulphate-reducing bacteria (SRB) which can generate H_2S at temperatures to 102°C (6) and pressures to 640 atmospheres (7), is compromised by suboptimal redox, pH and salinity. It cannot occur without a full substrate inventory and is susceptible to competitive microflora. Furthermore, the extent of the mixing zone in the water flooded region and the progress of the cold water front determine the dynamics of this environment.

Once H_2S has been generated downhole it can be irreversibly scavenged or reversibly scavenged; whatever remains will partition topsides between the fluid phases as dictated by the pressure, temperature, pH, salinity, chemistry and GOR of the system (1).

Inaccurate or missing values and other lacunae in the data sets needed for history matching or for making predictions, are a common feature with adverse effects upon the modelling process. In particular inaccuracy of the analytical technique for H_2S determination will adversely impact upon history matching. Predictions are further complicated by shut-in, the effects of multiple injection support and production from different sand layers.

Considering the complexity of the system, and the difficulties in obtaining a full and accurate data set, can reservoir souring realistically be predicted or are the values of variables so uncertain as to make the end result meaningless?

This non-prescriptive paper considers major criteria that need to be addressed for souring predictions. In particular, the difficulties in compiling a reliable data set, the problems of model building, and the effects of guesswork.

2 INTRODUCTION

Increasing H_2S production in oil well fluids over and above background, native, yields can result in the need to re-engineer produced fluid handling systems to avoid potential downstream health and safety and corrosion risks. Although the financial spin-offs from souring can be welcomed by the support industries, the costs borne by the operator often include loss of production through forced premature shut-in (4). A means of predicting the likelihood and extent of souring is therefore a potentially valuable tool to assist in the planning of a development. In the first instance of its evolution the predictive tool would be expected to match an existing sour gas profile where one was available. An accurate match might imply a corresponding accuracy of the predictive algorithms (4). However, given the known high variability in such profile measurements, coupled with possible unreliability of the reported values, too good a fit to such historical data might alternatively suggest a model with too much flexibility and an over-optimistic balance between bias and accuracy. Reducing one of these two loss assessments usually implies an inevitable increase in the other, when fitting models to real data for subsequent predictive purposes. The physical model also needs to offer the capacity to explain field observations and to make testable qualitative predictions. An accurate match to historical souring data in the absence of qualitative robustness in all circumstances is merely a numerical fit.

Fundamental to the development of a useful predictive tool are thus four cornerstones:

- The underlying conceptual model of the physical system
- The form in which the input data is derived
- The algorithms describing the conceptual model
- The accuracy of the predictions

The minimisation of uncertainties in both the conceptual model and input data must be the constant aim of the model builder towards achieving a credible and testable predictive algorithm.

3 CONCEPTUAL MODEL BUILDING

The range of models available to the oil industry include the prejudicial non-model: "will go sour/won't go sour, engineer to suit"; biogenic models based upon percentage conversion of breakthrough seawater sulphate to sulphide; emulators of varying degrees of simplicity; and simulators of varying degrees of complexity.

Indeed, we have found a very simple power law to give a surprisingly good fit to many historical H_2S profiles from seawater supported producers. Here gas phase H_2S concentration is related to time in days (post breakthrough) as follows:

$$[H_2S]_t = [H_2S]_T \, [t/\, T]^6$$

where T = time, since breakthrough, at which gas phase H_2S concentration was measured

t = time since breakthrough for which gas phase H_2S prediction is sought

$[H_2S]_T$ = known gas phase H_2S concentration at time T

$[H_2S]_t$ = predicted gas phase H_2S concentration at time t.

This is an easy to use rule of thumb method for the short term prediction of gas phase H_2S concentrations given recent historical data.

However, true conceptual model building requires a recognition of those parameters that ultimately influence the time to, and degree of, souring and the way in which these parameters interact. A conceptual model should address each of:

- Generation of H_2S downhole
- Transport of H_2S to the producer
- Partitioning of H_2S in reservoir fluids downhole and topsides

Complex models rely upon a wide range of data inputs and an initial judgement in the model building process has to be made upon the desired amount of complexity. With a biological H_2S generation model, the size, shape, location and activity of the three dimensional subterranean bioreactor dictates the overall production rate of sulphide. The bioreactor can be regarded either in terms of its overall ability to yield sulphide at the producer or as a web of exactly defined enzyme processes influenced by pressure, temperature, redox, pH, salinity and variable substrate limitations. A complex biological model in turn may require a matching mathematical model in processing the matrix scavenging effects before the bioreactor as a whole gives up its H_2S as it is transported in floodwater to the production well.

At the production wellhead, the produced fluids are directed to the first stage test separator. The concentration of H_2S in each of the fluid phases is dictated by the total mass of H_2S from the bottomhole environment, the pH, salinity and redox of the aqueous phase, the composition of the hydrocarbon phases, the gas/oil and water/oil ratios, production rates of the hydrocarbon phases and pressure and temperature at sampling.

Overlaying all the above is time. Bioreactors grow and mature, hydrocarbon production rates drop, and seawater injection rate, mixing zone size and seawater breakthrough increase over the lifetime of a development. Indeed, only at the end of a production well's life is the production profile finally known.

Against the backdrop of influential parameters, there are five forms of input data:

- Measured
- Assumed
- Calculated, to give assumed correct input
- Predicted
- History matched, to calibrate/fit the measured profile data

Based upon the 'choice' of any particular input value and the occasionally confused techniques of history matching and constant fitting, the careful selection of the form of input required in the predictive algorithm is that which gives confidence in the necessary complexity of the conceptual model and hence an appreciation of its limitations. For examples of preferred input forms, see Table 1.

TABLE 1

**Examples of input parameters, preferred and alternative,
necessary to attempt the prediction of oilfield reservoir souring.**

Parameter	Preferred form of input for use in predictive algorithm	Alternative form of input for use in predictive algorithm
Fluid phase H_2S concentrations	Measured	Gas phase measurement and balance calculated from assumed partition coefficients.
Production rates	Measured and predicted	None
GOR	Measured	None
Partition Coefficients	Calculated (from separator measurement conditions)	Calculated (under laboratory conditions) or assumed
Topsides pH	Measured	Calculated
Downhole pH	Measured (rare)	Calculated
Irreversible scavenging	History matched	Measured (under laboratory conditions)
Reversible scavenging	History matched	Measured (under laboratory conditions)
Seawater breakthrough (time)	Measured and predicted	None
Seawater breakthrough (cut)	Measured and predicted	None
Bottomhole pressure and temperature	Measured	Calculated

4 CHOICE OF FORM OF INPUT DATA

The best form is from direct and accurate measurements. In wells with native H_2S that have no secondary support 'souring predictions' based upon predicted production profiles are dependent primarily upon measured and calculated values. The accuracy of the prediction of the production profile is the greatest unknown and ever greater confidence in the H_2S prediction can be derived from careful analysis of the chemistry and flow rate of produced fluids and sampling conditions. Accurate data on the pH of the aqueous phase, and hence predominant moiety in which the sulphide exists, together with mass balance analyses from the separator enable a fix to be made on the partition coefficients k_o, k_w and k_{ow}. Downhole H_2S calculations are then based upon calculated pH and calculated partition coefficients at the pressures and temperatures of the system.

With secondary recovery, the biogenic and downhole scavenging models can greatly increase the number of input parameters required for a predictive model, however, there are still only three basic engineering questions:

- Can the reservoir support biogenic H_2S production?
- If yes, when?
- How much sulphide will be produced?

Biogenic H_2S production from viable mesophilic, thermophilic and hyperthermophilic (m-, t- and h-) SRB can only occur if the values of the following parameters are within the necessary boundary conditions to sustain the growth of these organisms:

- Temperature
- Pressure
- pH
- Redox potential
- Salinity
- Concentration of inhibitive species
- Concentration and rate of supply of sulphate
- Concentration and rate of supply of nutrients (in particular, the carbon source)

The temperature fundamentally dictates the ability of a reservoir to host viable organisms. Unless a reservoir is cool enough to support m-SRB, t-SRB and h-SRB then the boundary of the downhole bioreactor cannot be defined. The size, shape and location of the modelled bioreactor can be calculated from a physical cooling algorithm based upon assumed heat capacity of the water flooded zone and measured flood values and calculated downhole injection temperatures. History matching in cases of bottomhole cooling at the producer would enable a better fix on the downhole average rock heat capacity.

Within this thermally defined viability zone (or TVS, the Thermal Viability Shell), pressure, pH, redox potential and salinity affect the overall bioreactor efficiency whereas the availability of substrates set the ceiling for sulphate reduction. Two numbers are required to define bacterial activity within the bioreactor; the efficiency of sulphate reduction and the limit of sulphate that can be reduced. Both numbers constitute history matched data but could alternatively be derived from laboratory studies to limit the range of sulphate turnover to within feasible biological limits. Once the seawater flood has reduced the reservoir temperature to a suitably low level, SRB activity will (usually) ensue. The time to souring is then based upon the transport of H_2S from the bioreactor to the producer.

The question of when biogenic H_2S will appear and by how much are further affected by irreversible and reversible scavenging, but not presumably reservoirs where the matrix is already in equilibrium with native H_2S.

Irreversible scavenging is the capacity for the reservoir to react with generated H_2S, eg siderite scavenging:

$$FeCO_3 + H_2S \rightarrow FeS + CO_2 + H_2O$$

and the waterflooded zone will react with produced H_2S until saturated. This would cause a delay in the time to appearance of biogenic H_2S and a lower overall recovery of biogenic H_2S from the formation. Irreversible scavenging can thus be represented in a directly measurable dimension associated with time, together with predicted mass lost, rather than calculations based upon the assumed distribution of aqueous or mineral scavenger in the water flooded zone and the associated reaction kinetics dictated by the downhole thermal profile.

Reversible scavenging is the capacity for the swept zone to retard the transport of H_2S from the site of generation to producer, ie the formation will have a retention time somewhat akin to that observed in chromatography process. An adsorption/desorption kinetic model could be constructed using assumed inputs with constant fitting and history matching. This would achieve the same objective of increasing the time to appearance of H_2S at the wellhead by history matching biogenic H_2S output to the observed time to souring but with a handful of unprovable variables. A scavenging model based upon assumed values and history matching in conjunction with a biogenic model, similarly constructed, would thus have the maximum combination of uncertainties and therefore the capacity for exact matching even erroneous measured H_2S data. The more each parameter is related to reality, the less potential error there is likely to be. Here is where the 'engineering judgement' comes in!

5 PROBLEMS ASSOCIATED WITH THE CONSTRUCTION OF PREDICTIVE ALGORITHMS

This requires a system of equations to be written down so as to describe the joint behaviour of the principal variables as functions of time, distance and environment. There are two standard techniques. In one, the system is described by a set of differential equations and these are solved numerically by finite element techniques for the precise downhole conditions of each selected well. This is the usual format for an oilfield simulator: it is highly computer-intensive and typically requires a, possibly highly speculative, detailed knowledge of the oilfield geology downhole. In the other technique, a broad-brush approach is used to describe well conditions, enabling formal integration of the equations where appropriate, so as to produce a model which may appear algebraically complex, but is computationally easy to solve for specified conditions: which conditions are typically supplied as fairly broad, or even overall, well or oilfield average values, rather than point-by-point downhole values. Although this second technique, which is the one we currently utilize, can appear to involve crude compromises, it can be easier to manipulate, and hence understand. The resultant predictions may be as accurate as can realistically be hoped for with any predictor, given the extraordinary size and complexity of a typical reservoir-plus-wells system which is then coupled to the usual unpredictability of biological systems - which are here assumed to be the principal source of H_2S for sour wells.

The precursor to the above model for us was a program called Serec (standing for Secondary Recovery) which was based on a statistical calibration of data from a set of North Sea oilwells. This database had originally started out as a source of information on producing wells which had, or had not, gone sour. Statistical examination of this large database pointed strongly towards the incidental information supplied on the supporting injector(s) and hence confirmed the suggestion (as it could be described at the time) that there was a biological source for the H_2S, introduced, or stimulated, by the injection of seawater. This program also incorporated a 'siderite shield' effect which accounted for lower H_2S levels than otherwise expected in some fields. Chemical stripping was assumed to account for this observed statistical effect. This particular program was reported to have predicted that wells in the Gullfaks field would go sour long before any of them had in fact done so. However, whilst models which are predominately based on observed statistical correlations can be highly cost effective, they are unreliable outside the confines of the database which generated them, and are less amenable to manipulation for studying 'what-if' scenarios for field development.

Our overall approach in producing a mechanistic quantitative model for H_2S production has been to follow a typical litre of injected water as it moves through an idealised model of the reservoir, noting its temperature and pressure changes, and the implied consumption and dilution of its initial sulphate content in that particular reservoir's environment. The H_2S generated is then possibly retarded and/or absorbed by the surrounding geology before being partitioned between the various phases allowing, in particular, for the pH of the water and the presence, if necessary, of native quantities of H_2S in the oil and formation water.

The rate of consumption of sulphate by bacteria depends on pressure and temperature in a way which can be described by equations given in detail elsewhere (1). The temperature profile through a reservoir along the water path to a particular well changes dynamically with the passage of the cold-water front, which typically moves at a rate which is a small fraction of the speed of the injected water itself. This front eventually vanishes as a limit is reached to the amount of cooling which a given injector can accomplish when faced with the effectively infinite heat generating capacity of the Earth. The 'broad-brush' dimensionless equation we have used to describe the dynamics of this temperature profile (8) correctly predicted bottomhole cooling at a producer for Alwyn North before it occurred. (However the souring predicted has not occurred. We currently attribute this to extremely low bacterial efficiency for this particular injector/producer pair, rather than any nutritional deficiency or H_2S scavenging by the formation). The pressure falls off from the injector to the producer and we have modelled this, not unreasonably, by a simple square law. Hence we can write down an integral for the amount of sulphate consumed for our selected litre of water as it moves from injector to producer. The computer algebra package 'Maple V' was used to find this integral and the result was exported as about two pages of Fortran code, which was subsequently imported into a Quattro spreadsheet.

The laboratory derived rate parameter, as used by us for calculations of the consumption of sulphate, has been calibrated against observed consumption rates in the North Sea from measured H_2S in the production stream of various wells. We find that the downhole production efficiency for H_2S is around 10^{-5} times the laboratory derived value. This low efficiency is hardly surprising. A simple calculation (9) shows that just two cubic metres of porous rock - out of the tens of thousands typically flooded downhole - could otherwise support a highly active SRB population capable of generating all the H_2S actually observed in a biologically soured well. Our model also has a 'nutritional ceiling' parameter, which recognizes that not all the sulphate can be consumed due to restricted supplies of other necessary nutrients. This has also been calibrated against a selection of North Sea oilwells and we find a typical ceiling set at around 1% of the available 2650 mg/l of sulphate in North Sea seawater. Again, this low value is not surprising, since the consumption of 26.5 mg/l of sulphate, would imply 7 mg/l of H_2S in the oil phase at a typical partition ratio of 3:1 for oil:water - assuming the H_2S moiety in the aqueous phase is undissociated.

The range of temperatures over which the SRB are active determines the potential size of the TVS and hence the potential for production of H_2S. The extent to which this temperature range lies inside the injector/formation temperature range is the principal cause, we believe, of the typical delay seen between the onset of breakthrough and the start of souring. Geological adsorption/desorption mechanisms are not required in our model to explain this phenomenon, although we have allowed for absorption up to a specified limit for reservoirs with a siderite geology and no native H_2S, and an allowance for a diffusion delay to the H_2S on its travel through the formation rock. These two components are available for history matching purposes in our model, but we do not have sufficient information on these effects to use them for predictive purposes on new or proposed wells.

Partitioning of the H_2S between the various phases, once it has appeared, depends on the values of the various partition diffusion coefficients measured as partial pressure equivalents. There are computer programs available to calculate these coefficients given the chemical constituents of the phase under consideration. We have used a function fit to a table of values, calculated at various temperatures and pressures by such a program, for a typical North Sea production stream. One row of this table has been experimentally cross-checked using high-pressure autoclave equipment. The values were in broad agreement with each other, but not so close as to suggest that the theory is sufficiently well advanced to have removed the necessity for actual measurements with field samples.

To illustrate the variations in predicted gas phase H2S when using three of the techniques described above (power-law, Serec, TVS - in increasing degree of sophistication and presumed accuracy) the Ulay well (1) profile is given in Figure 1 and the corresponding set of predictions can be seen in Figure 2. A historical mid-point from each of the Serec and TVS curves is used as a basis for two corresponding power-law prediction curves. The Serec forecast appears to be over-predicting compared with the more sophisticated TVS model. The power-law curves appear quite reasonable as history-matchers/predictors, despite their remarkable simplicity.

6 CONCLUSIONS

There now appears to be general acceptance in the oil industry that souring occurs only after breakthrough and stems from downhole SRB activity. This acceptance is not confined to the North Sea environment, as can be seen for example in recent modelling procedures for an Alaskan field(3). Once H_2S has been produced and is free to dissolve in the various fluid partitions there is general agreement on the physical chemistry principles needed to predict its subsequent passage through the reservoir and associated production facilities. This understanding on its own can lead to successful history matching exercises for injector/producer well pairs, with the addition of so called 'fudge-factors', which have no explanatory power, to account for the actual rate of H_2S production by SRB activity. Further understanding of the downhole bioreactor dynamics, such as nutrient limiting conditions or siderite stripping, can replace these 'fudge-factors' with plausible explanatory causes in particular cases. Temperature is generally accepted as being important, although our model described above places more emphasis on this aspect than other models (3-5). Predictions for completely new fields remain fraught with difficulties, if for no other reason than the fact that forecasts of the all important production profile details, and in particular the crucial date of injection water breakthrough, may not be accurately known at the relevant time. Nevertheless, given a modicum of luck, plus sound engineering judgement concerning all the relevant parameters, predictions can be made and these will become increasingly more reliable for that well as historical data appears.

Figure 1:

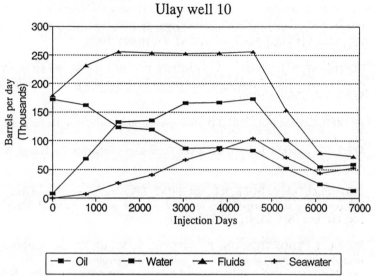

PRODUCTION PROFILE
Ulay well 10

Figure 2: Souring Profile predictions: Ulay well
H2S ppmv(gas) at selected Temp & Press.

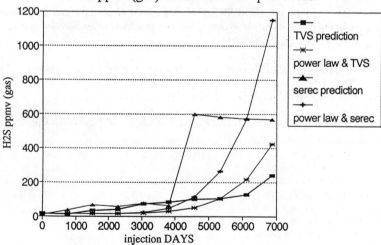

REFERENCES

1. EDEN, R.D., LAYCOCK, P.J. and FIELDER, M. *Oilfield Reservoir Souring.*
    UK Health and Safety Executive OTH 92 385, HMSO (1993).

2. COCHRANE, W.J., JONES, P.S., SANDERS, P.F., HOLT, D.M. and MOSLEY, M.J. *Studies on the thermophilic sulphate-reducing bacteria from a souring North Sea oilfield.*
    SPE 18368, (October 1988).

3. FRAZER, L.C. and BOLLING, J.D. *Hydrogen Sulphide Forecasting Techniques for the Kaparuk River Field.*
    SPE 22105, presented International Arctic Technology Conference, Anchorage, Alaska (May 29-31, 1991).

4. LIGTHELM, D.J., de BOER, R.B. and BRINT, J.F. *Reservoir souring: an analytical model for H_2S generation and transportation in an oil reservoir owing to bacterial activity.*
    Soc. Pet. Eng., SPE 23141, 369-378, (1991).

5. SUNDE, E., THORSTENSON, T., TORSVIK, T., VAAG, J.E. and ESPEDAL, M.S. *Field-related mathematical model to predict and reduce reservoir souring.*
    SPE 25197, presented SPE International Symposium on oilfield chemistry, New Orleans, La, USA,(March 2-5 1993).

6. STETTER, K.O., HUBER, R., BLÖCHL, M., KURR, M., EDEN, R.D., FIELDER, M., CASH, H. and VANCE, I. *Hyperthermophilic archaea are thriving in deep North Sea and Alaskan oil reservoirs.*
    Nature Vol. 365, 743-745, (1993)

7. STOTT, J.F.D. and HERBERT, B.N. *The effects of pressure and temperature on sulphate-reducing bacteria and the action of biocides in oilfield water injection systems.*
    J. Appl. Bac., 60, 57, (1986).

8. PLANTENKAMP, R.J. *Temperature distribution around water injectors: effects on injection performance.*
    Soc. Pet. Eng., SPE 13746, 513-521, (1985).

9. *Private communication*, FIELDER, M., VANCE, I. (BP Exploration Operating Company Ltd).

The Role of Sulphur and its Organic as well as Inorganic Compounds in Thermal Recovery of Oil

G. G. Hoffman, I. Steinfatt, and A. Strohschein

GERMAN PETROLEUM INSTITUTE, WALTER-NERNST.-STR. 7, D-38678 CLAUSTHAL-ZELLERFELD, GERMANY

Key words: Thermal reduction of sulphate, thermal recovery processes of crude oil, hydrogen sulphide, organic sulphur compounds, carbonic acids, carbon dioxide

1 Introduction

It is well-established in literature[1] that a great variety of chemical reactions exists involving elemental sulphur and its organic compounds in crude oil. This general feature of the manifoldness of reactions obviously has one of its origins in the variety of sulphur compounds being present in the crude oil and in the reservoir. Many different kinds of functional groups are conceivable, such as thiol groups, sulphide, disulphide and polysulphide groups, as well as thiophenic arrangements. Therefore, it is not surprising that one can find polar and apolar sulphur compounds with aromatic and non-aromatic nature in crude oil. The weak sulphur-sulphur and carbon-sulphur bonds relative to carbon-carbon bonds and the differencies in atomic radii, electronegativity, polarizability and other correlative conditions between the heteroelement sulphur and the carbon atom determine complicated concerted actions, and thus, the reactivity of these compounds. Moreover, a great variety of oxidation states of sulphur from -II to +VI in different organic and inorganic sulphur compounds is within easy reach for reactions. Thus, it is not surprising that elemental sulphur and its organic, as well as inorganic compounds are strongly participating in the chemistry of the crude oil during its geochemical genesis and during its production. All these features have to be seen together and contribute to the chameleon-like behaviour of sulphur chemistry in fossil fuels, which often seems to be unexpected and at first glance even unexplainable.

2 Scope

The scope of these investigations mainly has to be seen from the side of thermal recovery methods of crude oil. During these processes huge amounts of hydrogen sulphide (H_2S) may be produced together with carbon dioxide (CO_2) and small amounts of elemental hydrogen. In this connection the formation and reactivity of H_2S is of main interest, particularly with the prospect to come to a better understanding of the reactions and reaction pathways, which play a key role in the formation of H_2S and, in addition, of CO_2. At the injection wells temperatures may be as high as 320 °C. Temperatures at the producing wells can rise to 200 °C to 270 °C. During oil recovery the amount of H_2S may increase from 50 ppm up to 300 000 ppm and even more. Thus, enormous problems of corrosion and health security can arise.

In principle, formation of H_2S has always to be taken into account during oil and gas recovery. The origin of H_2S can have different reasons. Thus, it is obvious that H_2S can be formed by various processes, such as

1. microbial reduction of sulphate[2]
2. thermal decomposition of organic sulphur compounds, which are already present in crude oil and gas
   and
3. reactions of inorganic sulphur compounds, such as pyrite, pyrotite, elemental sulphur and sulphate.

At thermal recovery processes microbiological activity should be ruled out owing to high temperature. On the other side, thermal decomposition reactions of organic sulphur compounds (mainly of thiols and sulphides) under the formation of H_2S and olefinic compounds[3], as formulated for the thermal decomposition of organic disulphides in Eq. 1,

$$RCH_2SSCH_2R \xrightarrow{\quad T \quad} RCH=CHR + H_2S + 1/8\ S_8 \qquad (1)$$

have to be taken into account, anyway. If this reaction was the only source for H_2S-production during oil recovery with the help of thermal processes the amount of organically bound sulphur in the oil should decrease significantly over long periods (e.g. decades). However, it does not. The sulphur content of the crude oil remains almost constant. Therefore, H_2S-formation must result from other sources, in addition.

The formation of H_2S from inorganic magmatic sulphides, such as pyrite and pyrotite, is not of relevance at thermal recovery methods. These reactions have to be expected at much higher temperatures (about 1000 °C). On the other side, reactions of elemental sulphur with organic compounds are of importance. Elemental sulphur is often present in oil and gas bearing reservoirs. Moreover, the formation of elemental sulphur can be observed during reduction of sulphate, as well. Its reactivity against organic compounds leading to the formation of sulphur organic compounds is known[4-7]. At more elevated temperatures (600 °C and higher temperatures) these compounds decompose and H_2S together with olefinic compounds are formed[4].

Under certain conditions it should be paid great attention to H_2S-formation with respect to reduction of sulphate at high temperatures. Only a few original papers[8-13] are available concerning the redox reactions between sulphate and organic compounds in the presence of hydrogen sulphide as shown for methane in Eq. 2:

$$CH_4 + (NH_4)_2SO_4 \xrightarrow[\text{p,T}]{\text{H}_2\text{S (low conc.)}} CO_2 + (NH_4)_2S + 2\,H_2O \qquad (2)$$

It is reported[8,9] that ammonium sulphate reacts with organic compounds in the presence of at least small amounts of H_2S under redox conditions. H_2S serves as a catalyst. The reactions were run at temperatures in the range of 320 °C. The starting pressure at ambient temperatures was in the range of 23 bar.

The reaction of sulphate with either sulphide or H_2S is strongly pH-dependent. The oxidation potential of sulphate in the neutral pH-region is very low. Therefore, it is not astonishing that at atmospheric pressure and boiling temperatures of the inorganic and organic media (water, alkanes, alkyl substituted arenes) no reaction takes place within long reaction periods (e.g. hundreds of hours). However, the reaction may proceed very slowly over a very long time (e.g. geochemical time periods) and is indeed subject of discussion of geochemists[10].

Thiosulphate must be an intermediate during the reduction of sulphate. However, it is known from literature[14] that thiosulphate itself disproportionates, leading to the formation of sulphate and H_2S. This reaction already proceeds in neutral aqueous solution (and in the absence of elemental oxygen). The disproportionation is described to be quantitative. Nevertheless, small amounts of unreacted thiosulphate (about 5%) have been always recovered[14]. This fact shows clearly that the reaction must be an equilibrium reaction as formulated in Eq. 3.

$$SO_4^{2-} + H_2S \rightleftharpoons S_2O_3^{2-} + H_2O \qquad (3)$$

Under the conditions described in literature[14], the equilibrium in Eq. 3 is strongly shifted to the side of the educts. Therefore, the equilibrium in Eq. 3 should be influenced, for example by

1. the variation of the concentrations of the involved reactive species
or
2. side reactions with other components (for example organic compounds)

These and other influences have to be taken into account in oil and gas bearing reservoirs, and, in particular, during thermal recovery processes. They can participate in each other and thus increase their efficacy.

If, for example, thiosulphate is removed from the equilibrium in Eq. 3, which becomes possible by its reaction with organic compounds[15], the reduction of sulphate will easily proceed. Hence, thiosulphate can be formed again. The H_2S-formation is increasing because of decomposition reactions of the newly formed organic sulphur compounds. Consequently, an increasing concentration of H_2S in the equilibrium of Eq. 3 results. Furthermore, organic compounds are oxidized by a multiple step process to form carbonic acids. These acids themselves decompose at sufficient high temperatures to form CO_2 together with organic compounds, which have one methylen group less than the starting material. The overall reaction can be formulated as shown in Eqs. 4 and 5.

$$4\ RCH_3 + 3\ SO_4^{2-}\ \xrightarrow[p,T]{H_2S\ (low\ conc.)}\ 4\ RCOOH + 3\ S^{2-} + 4\ H_2O \qquad (4)$$

and

$$RCOOH \xrightarrow{high\ temperature} RH + CO_2 \qquad (5)$$

Moreover, anhydritic calcium sulphate serves as a sulphate reservoir. It slowly dissolves into the aqueous medium. Thus, at least steady state concentrations of sulphate in solution are maintained.

All the above mentioned concerted reactions contribute to a continuous formation and consumption of thiosulphate and, consequently, to increasing concentrations of H_2S.

Experimental

The experiments were conducted using glass cylinders, which were installed in stainless steel autoclaves. The starting pressure at ambient temperature was either the steel cylinder pressure of H_2S or Helium-pressure in the range of 14 bar, especially for those reactions, which were run in the absence of H_2S. The reactions were performed at temperatures between 200 °C and 360 °C, respectively (Tables II through V). After each run the organic layer was separated from the aqueous layer and the decrease of sulphate was determined by quantitative titration following the method of SCHÖNIGER[16]. The pH-value of the aqueous layer was determined after each run at room temperature with the help of a glass electrode. In the case of the reactions with crude oil an extraction of the organic layer with the help of liquid sulphur dioxide followed the autoclave reaction[17]. The organic layer was investigated with the help of gas chromatography. Gas chromatography was performed using a Hewlett Packard Model 5890 series II instrument, equipped with a Hewlett Packard Flame Ionization Detector (FID) and a Hewlett Packard Flame Photometric Detector (FPD). In Table I the analytical conditions are given. The aqueous layer was investigated with the help of high performance liquid chromatography comprising a high pressure pump Waters Model 510, a Machery-Nagel ET 250/8/4 Nucleosil 5 C_{18} column (particle size 5 μm) and a Waters 484 UV detector (260 nm). Flowrate: 0.6 ml/min. Mobile phase: methanol/$1.0 \cdot 10^{-3}$ M $NaH_2PO_4 \cdot H_2O$-solution (7:3, v/v). The method employed was developed by[18].

Table I. Analytical conditions

Injector temperature	(°C)	280
FID temperature	(°C)	330
FPD temperature	(°C)	280
GC column	DB 5 30 m x 0,252 mm film thickness 0,25 μm	
Carrier gas		Helium
Carrier gas flow	(ml/min)	1
Sample size	(μl)	1
Initial oven temperature	(°C)	35
Initial hold	(min)	5
Program rate	(°C/min)	5
Final oven temperature	(°C)	320
Final hold	(min)	10

Results and discussion

In Table II some reactions of sulphate are shown, which were performed in the absence of organic compounds. To rule out artifacts and to make sure that ammonium sulphate, as well as sodium sulphate do not react with the autoclave material, aqueous solutions of these compounds were treated at elevated temperatures in stainless steel autoclaves (Table II:1,2). It could be proved that no

Table II. Parameters of reactions without organic material

Nr.	sulphate [ml;mol/l]	S(-II) or S(0) [g]or[bar]*	pressure [bar]	temperature [°C]	time [h]	reduction [%]	pH-value	formation of
1	$(NH_4)_2SO_4$ 20;2	-	162	350	46	none	7	-
2	Na_2SO_4 20;1	-	118	290	24	none	7	-
3	$(NH_4)_2SO_4$ 20;1	Na_2S 1.25	124	325	43	none	10	-
4	Na_2SO_4 20;1	Na_2S 2.5	178	360	69	none	14	-
5	$(NH_4)_2SO_4$ 20;2	H_2S 17.1*	118	325	2	6	6	$S_2O_3^{2-}$, S_8
6	$(NH_4)_2SO_4$ 20;2	H_2S 16.4*	120	325	44	6	-	$S_2O_3^{2-}$, S_8
7	$(NH_4)_2SO_4$ 20;2	H_2S 19.1*	179	360	72	8	6	$S_2O_3^{2-}$, S_8
8	$(NH_4)_2SO_4$ 20;1	H_2S 22*	130	270	48	14	6	$S_2O_3^{2-}$, S_8
9	Na_2SO_4 20;1	H_2S 19*	126	325	47	6	8	-
10	$(NH_4)_2SO_4$ 20;1	S_8 1	130	270	72	2	3	H_2S
11	Na_2SO_4 20;1	S_8 1	130	270	72	none	3	H_2S

reduction of sulphate proceeds. The sulphate was recovered quantitatively.

Furthermore, it was necessary to prove the quantitative disproportionation of sulphur to sulphide and sulphate at high pH-values (pH = 10 to pH = 14) under the chosen conditions. It could be shown that no redox reaction takes place between aqueous solutions of sulphate and sulphide in the absence of organic compounds. The sulphate was recovered quantitatively (Table II:3,4).

On the other side, it could be demonstrated that aqueous solutions of sulphate react with H_2S under the same conditions. A reduction of sulphate in the range of 6% to 14% was found. In addition, small amounts of elemental sulphur could be identified in the resulting reaction mixtures (Table II:5-9). Surprisingly, sulphate reduction is increasing with decreasing temperature.

The reactivity of sulphate with elemental sulphur was investigated, in addition. For the chosen time and temperature program only a small effect in reduction of sulphate, but a rather high effect in H_2S-formation was found (Table II: 10,11). The formation of H_2S is not unusual, because elemental sulphur reacts under disproportionation to form H_2S and sulphate already under slightly alkaline conditions.

To demonstrate the redox reaction between aqueous sulphate solutions and organic compounds in the presence of H_2S as a function of temperature, toluene was used as a model compound (Table III). It could be clearly shown that sulphate does not react with toluene at 270 °C in the absence of H_2S. The sulphate was recovered to 100% (Table III:1). On the other side, in the presence of H_2S a sulphate reduction in the range of 4% to 95% was found, depending on the reaction conditions (Table III:2-8). The dependence on the reaction time can be estimated by comparing the results of Table III:2 with those of Table III:3. The dependence on the concentration of H_2S is expressed in the results of Table III:6 compared to those of Table III:7. The reduction in the presence of high concentrations of H_2S is slow at temperatures between 200 °C and 220 °C (Table III:2-4), but accelerates markedly with increasing temperature (Table III:5,6 and 8). In the aqueous layer of all runs, benzoic acid could be detected with the help of high performance liquid chromatography and comparison with authentic samples.

In another reaction it could be shown that sulphate obviously reacts with elemental sulphur under redox conditions, as well (Table III:9). The formation of H_2S, as described in Table II:10,11, could be the driving force and may serve as a sufficient explanation.

Gas chromatographic investigations of the resulting reaction mixtures showed that different organic sulphur compounds are formed. Some of them are identified.

Table III. Parameters of the reactions in the presence of toluene

Nr.	Fig. FPD	sulphate [ml;mol/l]	gas [bar]	pressure [bar]	temperature [°C]	time [h]	reduction [%]	pH-value (after reaction)	formation of
1		$(NH_4)_2SO_4$ 20;1	He 12,8	-	270	76	none	7	-
2		$(NH_4)_2SO_4$ 20;1	H_2S 20	48,9	200	82	none	4,6	benzoic acid (traces)
3		$(NH_4)_2SO_4$ 20;1	H_2S 20	49,2	200	528	4,1	5,9	CO_2, S_8
4	1	$(NH_4)_2SO_4$ 20;1	H_2S 20	52,7	220	80	4,2	6,7	benzoic acid CO_2
5	2	$(NH_4)_2SO_4$ 20;1	H_2S 20	81,4	250	80	13,3	7,1	benzoic acid CO_2
6	3	$(NH_4)_2SO_4$ 20;2	H_2S 18,8	119,0	270	80	37,2	7,7	benzoic acid
7		$(NH_4)_2SO_4$ 10;2	H_2S 10;0,1+	114	270	72	none	6	-
8	4	$(NH_4)_2SO_4$ 20;1	H_2S 19.5	168	325	53	95	6	benzoic acid CO_2, benzene
9		$(NH_4)_2SO_4$ 20;1	S_8 1*		270	46	49,2	8,5	benzoic acid CO_2, H_2S

+ (ml;mol/l); * (g)

The investigations show a strong temperature dependence of the formation of the different species. At 200 °C small amounts of α-toluene thiol (1) and elemental sulphur (2) are formed. These compounds could be detected as the main products. The reaction still proceeds at 220 °C and the amount of α-toluene thiol increases (Figure 1). At 250 °C dibenzyl disulphide (3) is formed, in addition (Figure 2). At 270 °C benzyl sulphide (4) is produced together with other sulphur organic compounds (Figure 3). At 325 °C the amounts of α-toluene thiol (1), elemental sulphur (2), and dibenzyl disulphide (3) decrease drastically and other sulphur containing species (benzothiophens, dibenzothiophens) are formed instead (Figure 4).

The results of the redox reactions of ammonium sulphate, as well as sodium sulphate and H_2S in the presence of crude oil are summarized in Tables IV and V. The reactions were performed at three temperature windows, namely at 250 °C, 270 °C, and 320 °C, respectively. Sulphate reduction was found, reaching from 5% to 98%, depending on the reaction conditions. The reduction of sulphate is more time consuming if sodium sulphate is used instead of ammonium sulphate. E.g., the reduction of $(NH_4)_2SO_4$ at 320 °C is quantitative after 72 h, whereas for the same amount of Na_2SO_4 more than 500 h are necessary (Table IV:9; Table V:8). Only about one third of sodium sulphate is reduced at 320 °C after 300 h (Table V:7). Moreover, the reduction of sulphate at lower temperatures is slow. E.g., after 72 h only about 5% of total ammonium sulphate is reduced between 250 °C and 270 °C, respectively (Table IV:7,8).

At 320 °C the reduction of sulphate even could be achieved in the absence of H_2S (Tables IV and V:6). This effect easily can be explained by the formation of H_2S caused by decomposition reactions of organic sulphur compounds, which are already present in the crude oil. However, for the investigated time intervall, the reaction does not proceed significantly at 270 °C (Tables IV and V:5). For comparative studies, crude oil was treated in the presence of water and water/H_2S, as well (Tables IV and V:2-4). The results of those reactions, performed in the presence of water, clearly show that H_2S is formed at 320 °C (Tables IV and V:2).

After each run the recovered crude oil was extracted with the help of liquid sulphur dioxide. The extracts were investigated with the help of gas chromatography. The chromatograms were compared with that one of the SO_2-extract of the thermally untreated, but at ambient temperature vacuum stripped authentic sample of the crude oil (Tables IV and V:1). Figures 5 through 10 show some of the FPD chromatograms. From these chromatograms the product distribution of the sulphur compounds, deriving from the reactions of sulphate in the presence of crude oil (Figure 6-10) can be evaluated and compared with the distribution of the sulphur compounds of the extract of the unreacted crude oil (Figure 5). It

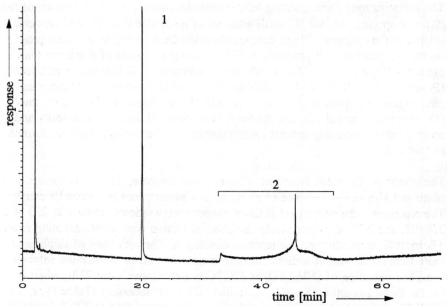

Figure 1. FPD chromatogram of the 220 °C-reaction of $(NH_4)_2SO_4/H_2S/toluene$

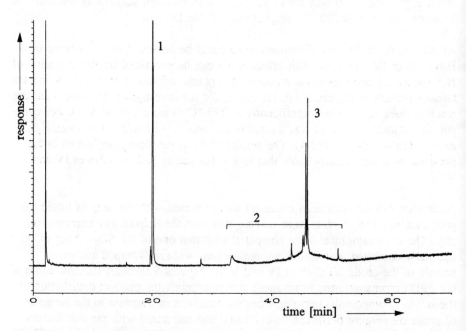

Figure 2. FPD chromatogram of the 250 °C-reaction of $(NH_4)_2SO_4/H_2S/toluene$

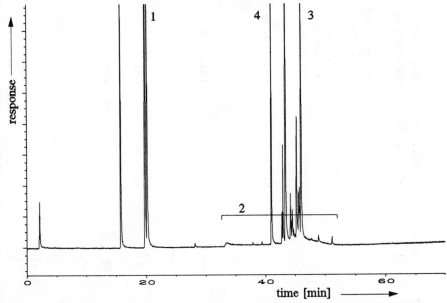

Figure 3. FPD chromatogram of the 270 °C-reaction of $(NH_4)_2SO_4/H_2S/$toluene

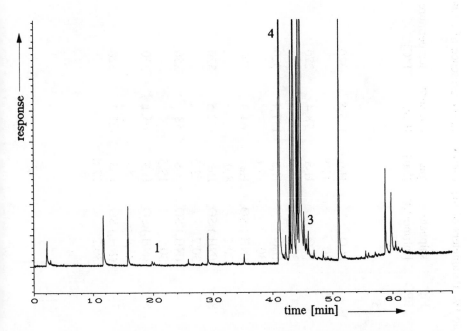

Figure 4. FPD chromatogram of the 325 °C-reaction of $(NH_4)_2SO_4/H_2S/$toluene

Table IV. Parameters of the reactions in the presence of crude oil A

Nr.	Fig. FPD	sulphate [ml;mol/l]	gas [bar]	pressure [bar]	temperature [°C]	time [h]	reduction [%]	pH-value (after reaction)	comment*/ formation of
1	5	-	-	-	-	-	-	-	SO_2-extracted crude oil*
2		H_2O 30	He 11,9	135,8	320	72	-	7,8	CO_2 (traces) H_2S (traces)
3		H_2O 30	H_2S 18,1	78,4	270	72	-	4,0	CO_2 (traces)
4		H_2O 30	H_2S 17,6	137,1	320	72	-	4,0	CO_2 (traces)
5		$(NH_4)_2SO_4$ 20;2	He 8,0	64,9	270	80	none	3,7	-
6	8	$(NH_4)_2SO_4$ 30;2	He 12,5	182,5	320	68	44,3	9,3	CO_2, H_2S
7		$(NH_4)_2SO_4$ 30;2	H_2S 18,3	58	250	72	5,9	7,3	CO_2
8	6	$(NH_4)_2SO_4$ 25;4,6	H_2S 18,6	65,1	270	72	5	7,5	CO_2
9	7	$(NH_4)_2SO_4$ 30;2	H_2S 17,8	187,1	320	72	98	9,3	CO_2

Table V. Parameters of the reactions in the presence of crude oil A

Nr.	Fig. FPD	sulphate [ml; mol/l]	gas [bar]	pressure [bar]	temperature [°C]	time [h]	reduction [%]	pH-value (after reaction)	comment*/ formation of SO₂-extracted crude oil*
1	5	-	-	-	-	-	-	-	SO₂-extracted crude oil*
2		H_2O 30	He 11,9	135,8	320	72	-	7,8	CO_2 (traces) H_2S (traces)
3		H_2O 30	H_2S 18,1	78,4	270	72	-	4,0	CO_2 (traces)
4		H_2O 30	H_2S 7,6	137,1	320	72	-	4,0	CO_2 (traces)
5		Na_2SO_4 20;1	He 13,9	76,4	270	82	none	5,0	CO_2
6		Na_2SO_4 30;2	He 13,9	148,7	320	500	36,2	7,6	CO_2, H_2S
7	9	Na_2SO_4 30;2	H_2S 18,3	137,8	320	295	30,5	8,2	-
8	10	Na_2SO_4 30;2	H_2S 17,6	139,4	320	529	98,6	7,5	CO_2

easily can be seen that qualitatively and quantitatively the most impressing alterations are found for those reactions performed at 320 °C. The alterations of the extractable organic sulphur compounds are more pronounced for those reactions, which were run in the presence of sodium sulphate and for longer reaction times. In the presence of ammonium ions additional sulphur containing organic compounds are formed compared to the reactions in the presence of sodium sulphate. An explanation for this behaviour may be given with the Willgerodt-Kindler reaction[19], which may be involved in the formation of these compounds.

Conclusions

In a series of experiments the thermal reduction of sulphate could be proved in the absence and in presence of toluene as a model compound and of crude oil. The results clearly show that

- the reduction of sulphate in the presence of hydrogen sulphide will proceed within a reasonable reaction time even in a pH-region between pH 5 and pH 9 if the employed temperatures are sufficient high.
- H_2S serves as a catalyst and cumulates during the reaction. High concentrations of H_2S accelerate the reduction process significantly.
- in the presence of ammonium ions the reaction proceeds more frequently. Additional organic sulphur compounds are formed compared to the reactions in the presence of sodium sulphate. The reaction mechanism playing the key role for this behaviour, may be traced back to the Willgerodt-Kindler reaction, but still is tentative.
- high temperatures accelerate the redox reaction. Nevertheless, there is evidence that the reaction proceeds at even lower temperatures, however, over longer reaction periods.
- in the presence of organic compounds the reduction of sulphate increases drastically. The formation of additional oxygen-, as well as sulphur-containing organic species is observed.
- the formed organic acids and the sulphur-containing compounds decompose themselves at sufficient high temperatures to form H_2S and CO_2, which strongly contribute to the rising H_2S- and CO_2-concentrations during thermal recovery processes.

In the case of crude oil as reactant initial H_2S is not necessary to start the reduction of sulphate. This result is certainly the most interesting. An explanation may be given with the organic sulphur compounds already present in crude oil. Thermal decomposition of these compounds leads to the formation of H_2S, which for its part starts the reduction of sulphate. Therefore, quantitative reduction of sulphate in the absence of H_2S requires longer reaction periods.

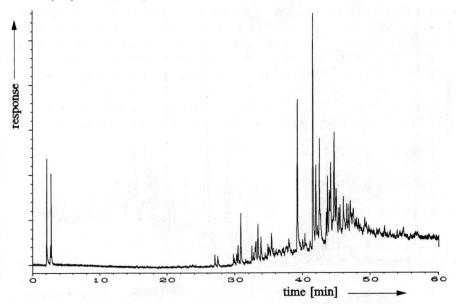

Figure 5. FPD chromatogram of the SO_2-extract of crude oil A

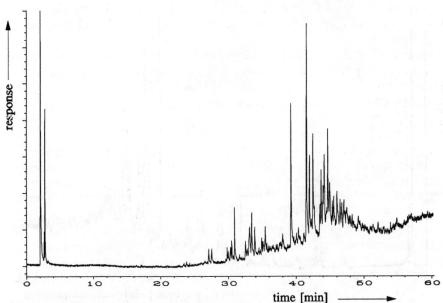

Figure 6. FPD chromatogram of the SO_2-extract of the 270 °C-reaction of $(NH_4)_2SO_4/H_2S/$crude oil A

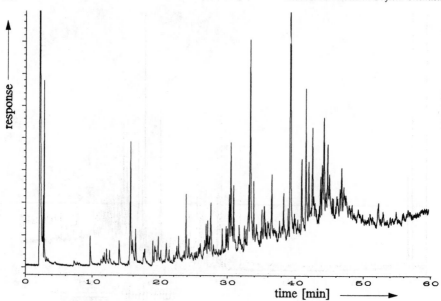

Figure 7. FPD chromatogram of the SO$_2$-extract of the 320 °C-reaction of (NH$_4$)$_2$SO$_4$/H$_2$S/crude oil A

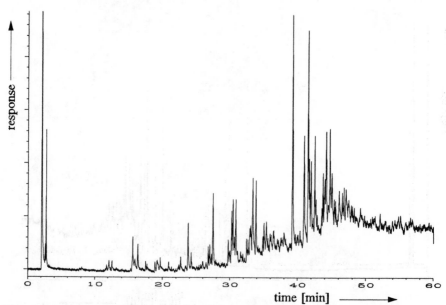

Figure 8. FPD chromatogram of the SO$_2$-extract of the 320 °C-reaction of (NH$_4$)$_2$SO$_4$/crude oil A

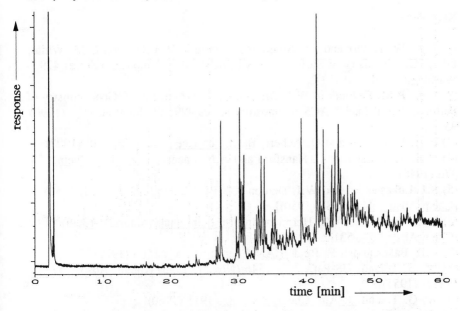

Figure 9. FPD chromatogram of the SO_2-extract of the 320 °C-reaction of Na_2SO_4/H_2S/crude oil A; reaction time 295 h; 30,5% reduction of SO_4^{2-}

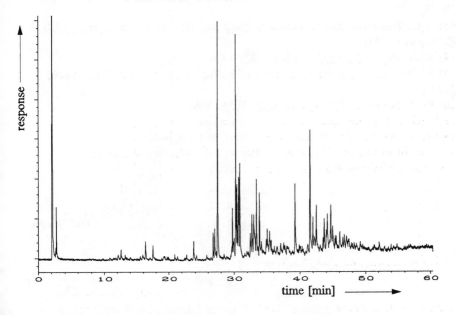

Figure 10. FPD chromatogram of the SO_2-extract of the 320 °C-reaction of

References

(1) E.g.: W. L. Orr and J.S. Sinninghe Damsté in W.L.Orr and C.M. White, Eds., "Geochemistry of Sulfur in Fossil Fuels" ACS Symposium Series 429, Washington, DC, (1990) 2.

(2) E.g.: P.M. Fedorak in W.L.Orr and C.M. White, Eds., "Geochemistry of Sulfur in Fossil Fuels" ACS Symposium Series 429, Washington, DC, (1990) 93.

(3) E.g.: E. Fromm and O. Achert, Ber.Dtsch.Chem.Ges., 36, 538 (1903).

(4) H.E. Rasmussen, R.C. Hansford and A.N. Sachanen, Ind.Eng.Chem., 38, 376 (1946).

(5) S.J. Lukasiewicz and W.I. Denton, US-Pat., 2,515,928 (July 18, 1950) [Chem.Abstr., 44, 9668b (1950)].

(6) F.G. Vigide and A.L. Hermida, Inform.quim.anal. (Madrid), 13, 61 (1959) [Chem.Abstr., 54, 5314c (1960)].

(7) R.B. Baker and E.E. Reid, J.Am.Chem.Soc., 51, 1566 (1929).

(8) W. G. Toland, US-Pat., 2,722,546 (Nov. 1, 1955) [Chem. Abstr., 57, 11111i (1955)].

(9) W. G. Toland, J. Am. Chem. Soc., 82, 1911 (1960).

(10) W. L. Orr, Am.Ass.Pet.Geol.Bull., 58, 2295 (1974).

(11) W. L. Orr, Geol.Soc.Am.Abstr.Progm., 14, 580 (1982).

(12) I. Steinfatt and G.G. Hoffmann, Phosphorus, Sulfur, and Silicon, 74, 431 (1993).

(13) G.G. Hoffmann and I. Steinfatt, Preprints, Div. of Petrol.Chem., ACS, 38(1), 181 (1993).

(14) W.A. Pryor, J.Am.Chem.Soc., 82, 4794 (1960).

(15) S. Oae, Ed. "Organic Chemistry of Sulfur" Plenum Press, New York, (1977).

(16) W. Schöniger, Mikrochim. Acta, 1956, 869.

(17) G.G. Hoffmann, manuscript in preparation.

(18) M. S. Akhlaq, private communication, to be published.

(19) See for example: (15) or W.A. Pryor, Ed. "Mechanisms of Sulfur Reactions", McGraw-Hill, New York, (1962).

The Development, Chemistry and Applications of a Chelated Iron, Hydrogen Sulphide Removal Process

D. McManus

ARI TECHNOLOGIES INC, 1950 S. BATAVIA AVENUE, GENEVA,
ILLINOIS 60134, USA

A. E. Martell

DEPARTMENT OF CHEMISTRY, TEXAS A & M UNIVERSITY, COLLEGE
STATION, TEXAS 77843, USA

The industrial use of iron as a regenerable
oxidant for the conversion of hydrogen sulphide
to elemental sulphur can be traced back to dry
oxidation processes employing hydrated iron(III)
oxide which absorbs H_2S forming ferric sulphide
according to:

$$2\ Fe_2O_3 + 6H_2S\ \ =\ \ \ \ \ 2\ Fe_2S_3 + 6\ H_2O$$

Regeneration was achieved with the concurrent
formation of sulphur simply by exposure to
atmospheric oxygen:

$$2\ Fe_2S_3 + 3O_2\ \ \ +\ \ \ \ \ \ \ 2\ Fe_2O_3\ + 6\ S$$

This process was quite efficient but progressive
accumulation of sulphur eventually expended the
iron oxide reactant.

Subsequently, hydrated iron (III) oxide (i.e.
$Fe(OH)_3$) suspensions in aqueous, alkali metal
carbonate solution were introduced, for example,
the Ferrox and Manchester processes. Although H_2S
absorption efficiencies were satisfactory, the
difficulty in separating sulphur product from the
iron reactant endured.

Hexacyanoferrate based processes were developed and enjoyed limited applications around mid-century. These may have been invented following observations of Prussian blue formation in ferric oxide processes that treated gas streams containing both H_2S and HCN such as coke oven gas. Proposed reactions for H_2S oxidation and regeneration of the iron containing oxidant by aeration are, respectively.

$$2H_2S+Fe_4[Fe(CN)_6]_3+2Na_2CO_3=2Fe_2[Fe(CN)_6]+Na_4[Fe(CN)_6]+2H_2O+CO_2+2S$$

and

$$2Fe_2[Fe(CN)_6]+Na_4[Fe(CN)_6]+O_2+H_2CO_3=Fe_4[Fe(CN)_6]_3+Na_2CO_3+2H_2O$$

It is noted that these equations, from Gas Purification by Kohl and Riesenfeld, indicate the presence of CO_2 and carbonic acid. In practice, these compounds would be converted to carbonate/bicarbonate at the alkaline pH of the process.

The Autopurification and the Staatsmijnen-Otto processes employed suspensions of complex ferriferrocyanide compounds in aqueous, alkaline solution and suspension and accomplished regeneration with air. Process chemistry equations are likely similar to those given above.

The Fisher process used alkaline ferricyanide solution to oxidize H_2S to sulphur according to:

$$2K_3[Fe(CN)_6]+K_2CO_3+H_2S=2K_4[Fe(CN)_6]+S+2KHCO_3$$

Regeneration proved difficult but was achieved electrolytically:

$$2K_4[Fe(CN)_6]+2KHCO_3=2K_3[Fe(CN)_6]+H_2+2K_2CO_3$$

The introduction of chelated iron processes commenced in the early 1960's with the assignment of a significant patent to Humphreys and Glasgow (1). No successful commercial process emerged however, and it later became apparent that extremely rapid in-process degradation of the amino polycarboxylate ligand precluded broad commercialization.

Research on iron based, H_2S oxidation processes continued and eventually led to the introduction of current state of the art chelated iron processes in which the iron is held in true aqueous solution at mildly alkaline pH values by multidentate, amino polycarboxylate ligands such as EDTA, HEDTA and NTA.

A superior chelated iron catalyst, stable against hydrated iron(III) oxide precipitation throughout the entire pH range, incorporates a hexitol to augment the amino polycarboxylate (2).

Process chemistry reactions are:

Absorption: $H_2S(g)$ $\rightleftharpoons$ $H_2S(aq)$

Ionization: $H_2S(aq)$ $\rightleftharpoons$ $H^+ + HS^-$

Oxidation: $2Fe^{3+}L + HS^-$ $\rightleftharpoons$ $2Fe^{2+}L + H^+ + S$

Regeneration: $2Fe^{2+}L + H_2O + \tfrac{1}{2}O_2$ $\rightarrow$ $2Fe^{3+}L + 2\ OH^-$

Overall: $H_2S + \tfrac{1}{2}O_2$ $\rightarrow$ $H_2O + S$

Small capacity systems (3) commissioned in the 1970's discarded substantial amounts of chelated iron catalyst solution with the sulphur product as a wet cake and so masked the loss of chelons by chemical degradation.

It was not until the first large plants (4) equipped with sulfur melters and essentially operating with a captive catalyst inventory were brought on line around 1980, that the problem of chelon degradation was fully recognized.

An urgent research program was initiated with the objective of firstly defining the underlying chemical changes responsible for the unacceptable operating costs and secondly, to derive a way to correct the situation.

Liquid chromatographic procedures (5),
appropriately modified (6), were used to monitor
the concentration of chelating agents with
respect to time in a laboratory scale process
simulation reactor.

Periodically withdrawn samples of the iron
chelate solution were prepared for reverse phase,
ion-paired, liquid chromatography by
quantitatively replacing the iron(III) with
copper(II) to produce a negatively charged
complex that would form an ion-pair with the
quaternary ammonium counterion present in the
mobile phase. The interaction of this ion pair
with the octadecyl functionality of the reversed
stationary phase allowed baseline separation of
most common amino polycarboxylates in a
relatively short time, Figure 1.

By this means, aqueous solutions of iron chelated
with the common amino polycarboxylates such as
HEDTA, EDTA, NTA, DTPA, IDA and CDTA (7) were
demonstrated to undergo rapid degradation during
redox cycling, Figures 2a & b illustrate
chromatograms from fresh and used catalyst
solutions, respectively.

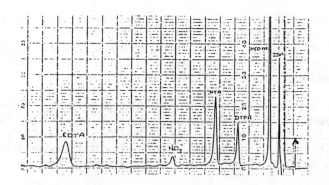

Figure 1
Resolution of mixed chelating agents

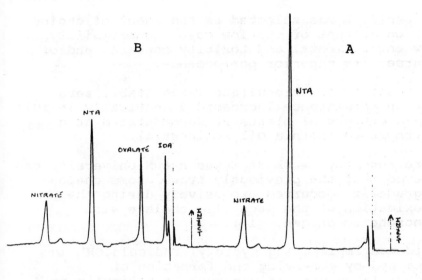

Figure 2
Chromatograms of fresh (A) and used (B) chelated iron catalyst solutions

Liquid chromatography also provided information on the degradation products. With NTA for example, two new peaks appeared in chromatograms from used solutions. Tentatively assigned identities based on retention times suggested that these products were IDA and oxalate.

Subsequent gas chromatographic analysis (8) of the butylated and trifluoroacetylated derivatives of the degradation products and pure reference compounds confirmed this.

With the knowledge that chemical degradation was responsible for the unacceptable operating costs, all efforts focused on finding effective stabilizers.

Additives such as antioxidants, buffers, sacrificial agents and free radical scavengers were screened in a series of 24 hour tests. Several, such as thiosulfate, t-butanol and thiocyanate, emerged as effective.

Thiosulfate was selected as the agent of choice
(9) on account of its low cost, compatibility,
low environmental and toxicity concerns and of
course, its superior performance.

By addition of thiosulfate based stabilizers
chelon persistence increased dramatically in full
scale commercial plants as demonstrated at a
Southern California Oil Refinery(4).

More recently, work at Texas A & M University has
proved that the previously troublesome chelon
degradation occurred exclusively during the
reoxidation of the iron (II) chelate with
atmospheric oxygen, (10).

Evidence implicating hydroxyl radicals, OH·, was
obtained by observing the formation of
hydroxylated derivatives such as salicylic acid
from added benzoic acid. Subsequently, ESR
spectroscopy employing spin trapping reagents has
strongly implicated free hydroxyl radicals,
generated by Fenton type chemistry, as the
oxidant responsible for chelon degradation,
Figure 3.

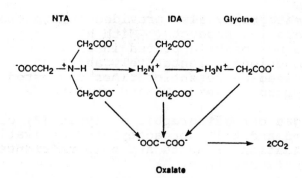

Figure 3
Proposed degradation scheme for NTA.

$$2 \ FeNTA^- + O_2 + 2H_2O \ \rightarrow \ 2 \ FeNTA + H_2O_2 + 2 \ OH^-$$

$$FeNTA^- + H_2O_2 \ \rightarrow \ FeNTA + OH^- + OH\bullet$$

Addition of the enzyme catalase, which destroys hydrogen peroxide, was shown to increase NTA stability in a laboratory process reactor.

With expensive chelon loss greatly reduced and under control, the number of commercial plants grew rapidly as the processes cost effective capacity rose to roughly 10 tons sulphur per day, an amount still limited by chelon loss and Claus process relative economics.

Plant configurations for chelated iron processes can be many and varied but all share the basic vessels and unit operations.

Figure 4 illustrates the process flow for a conventional plant where the sweetened gas is kept separate from the regeneration air.

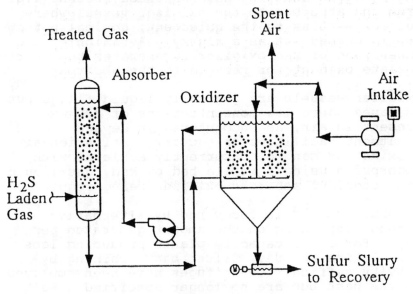

Figure 4
Conventional Plant

Sour gas is contacted with the oxidized chelated iron circulating catalyst solution in an absorber vessel where the H_2S dissolves, ionizes primarily to HS^- and is then oxidized to elemental sulphur by the iron(III) chelate which is reduced to the iron (II) chelate.

Chelated iron reaction kinetics are greatly improved relative to earlier iron based processes and exceed those of competing vanadium based processes by such a margin to enable elimination of delay tanks downstream of the absorbers as required to complete sulphide oxidation.

A considerable variety of gas-liquid contactors have found application in chelated iron installations. Selection criteria depend on the composition, pressure and volumetric flow rate of the sour gas stream. Types of absorbers include liquid filled columns; packed towers with a variety of packings, both fixed and mobile; static or pipeline mixers; venturis or eductors and spray chambers. Combinations of absorbers have been used.

The catalyst solution and suspended sulphur flow from the absorber to the oxidizer vessel where sulphur settles to the quiescent, conical bottom and is pumped out as a slurry. Aeration of the upper zone of the oxidizer regenerates the ferric chelate oxidant for recycle to the absorber.

Oxidizer vessels are generally liquid filled but in cases where low concentrations of H_2S are present in air, as encountered in sewage treatment facilities or factory ventilation air, reoxidation can be concurrently achieved with absorption using a mobile bed packed, gas-liquid contactor, no separate oxidizer being required.

Sulphur product is recovered by alternative methods depending on the amount generated per day. For small capacity plants producing less than 3 tons per day, filter bags draining by gravity suffice. Centrifuges have been employed in the past but are no longer specified. Belt filters with filtrate return and subsequent water washing are used for larger capacity plants. The washed sulphur is reslurried with pure water and fed to a sulphur melter operating under pressure at about 135°C. Molten sulfur disengages from the catalyst solution in a liquid-liquid separator and is withdrawn to storage for sale or disposal.

Some side reactions occur and up to a point can be valuable. These include sulphur oxo-acid by-product salt formation caused by air induced oxidation of unreacted dissolved sulphide. Thiosulphate and its oxidation products tetrathionate and sulphate are produced.

$$2 \ HS^- + 2O_2 \quad \rightarrow \quad S_2O_3^{2-} + H_2O$$

$$2S_2O_3^{2-} + \tfrac{1}{2}O_2 + H_2O \quad \rightarrow \quad S_4O_6^{2-} + 2 \ OH^-$$

$$S_2O_3^{2-} + 2O_2 + H_2O \quad \rightarrow \quad 2SO_4^{2-} + 2H^+$$

By-product thiosulfate so formed is of use as a free radical scavenger but higher concentrations provide a diminishing effect. Irrevocable discard of catalyst solution is required when by-product salt concentrations reach about 200 g/L.

An innovative, earlier, chelated iron process design, Figure 5, finding extensive service in amine acid gas treating, integrated the absorber

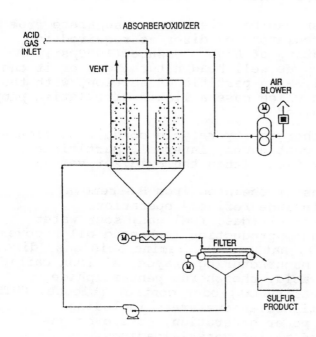

Figure 5
Autocirculation Unit

and oxidizer into a cone bottomed, cylindrical
single vessel. The oxidizer zone occupied the
outer annulus and the absorber consisted of a
centrally disposed open ended, submerged
cylinder. Lift generated by aeration achieved
torroidal liquid circulation without the use of
pumps. The sweetened process gas having no value
was directly discharged to atmosphere with the
spent oxidation air. This design allowed some HS^-
to react with dissolved oxygen to form
thiosulfate in amounts roughly equal to about 6%
of the sulfur input as H_2S. However, plant
capacities were generally small, around 5 tons of
sulfur per day so catalyst discard was tolerable
especially at the 10mM iron chelate
concentrations employed.

More recent process designs (11), Figures 6 and
7, have achieved considerable success in reducing
by-product salt formation by ratio controlled
mixing of iron(III) chelate solution with
absorber effluent so as to complete sulphide
oxidation before dissolved oxygen is encountered
in an autocirculatory multizone vessel.

The absorber can be external and separate from
the autocirculatory oxidizer as required for
direct treating of high pressure (1000psi)
streams such as well head natural gas or it can
be integral, as a partitioned section, with the
oxidizer in which case a liquid circulation pump
is not required.

Note that the gravity recycle conduit of figure 6
represents continuous, internal, partitioned,
circulatory flow within the oxidizer vessel.

Applications of chelated iron H_2S removal
processes include refinery operations
(hydrotreater off-gas, fuel gas, sour water
stripper gas), production (enhanced oil recovery,
CO_2 recycle), natural gas (amine acid gas, direct
treating), manufacturing (rayon, silicon carbide,
phosphoric acid, phosphorus pentasulphide,
beverage quality CO_2) odor control (biogas, WWTP)
and other miscellaneous areas including
geothermal power generation, coke oven gas
treating and marine vessel loading.

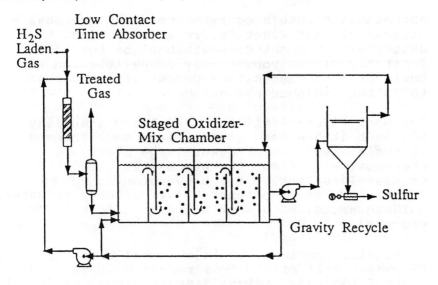

Figure 6

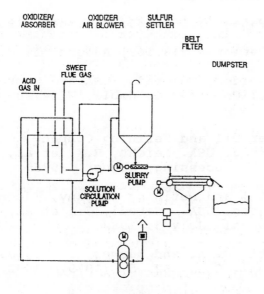

Figure 7

Advantages of chelated iron processes are ease of operation, fast kinetics, relatively high selectivity, no gas compositional or turn down limitations, environmentally compatible and non-toxic process chemicals and cost effectiveness up to 10 tons sulphur per day.

The most severe limitation is chelon stability. Although it has been greatly improved, the cost of chelon replacement makes large plants uneconomical. Also, the process is not completely selective for conversion of H_2S to sulfur. Up to 2% of the sulfur converts to water soluble salts which must be removed from the process.

As studies continue, it is expected that capacity increases will result from research aimed at controlling free radical damage to the organic chelating agents and that process design modifications will further lower the yield of by-product salts.

References & Notations

(1)   U. S. Patent 3,068,065; December, 1962.

(2)   U. S. Patent 4,218,342; August, 1980.

(3)   Amax, Golden, Colorado, USA and Hooker Chemical Company, Columbia, Mississippi, USA.

(4)   Fletcher Oil and Refining Company, Carson, California, USA. Leicht, R.K. et al, Chemical Processing, August, 1986.

       U. S. Oil and Refining Company, Tacoma, Washington, USA. Cabodi, J., et al, Oil and Gas Journal, July 5, 1982.

(5)   Perfetti, G. A. and Warner, C. R.; J. Assoc. Off. Anal. Chem. 62, 5, (1979), 1092.

(6)   McManus, D., 41st Pittsburg Conference and Exposition on Analytical Chemistry and Applied Spectroscopy, New York City, 1990.

(7)   HEDTA   =   Hydroxyethyl ethylenediamine
                         triacetic acid.

       EDTA   =   Ethylenediamine tetraacetic
                         acid.

       NTA   =   Nitrilotriacetic acid.

       DTPA   =   Diethylenetriamine pentacetic
                         acid.

       IDA   =   Iminodiacetic acid.

       CDTA   =   Cyclohexanediamine
                         tetraacetic acid.

(8)   Warren, C. B. and Malec, E. J.; J.
       Chromatography, 64, (1972), 219-237.

(9)   U. S. Patent 4,622,212; November, 1986.

(10)   Chen, D., et al. Can. J. Chem., Vol. 71,
       1993.

(11)   Hardison, L.C., et al, ASME Petroleum
       Division Energy Sources Technology
       Conference; January, 1992.

       Hardison, L. C., AIChE, National Spring
       Meeting, New Orleans, Louisiana, USA;
       March, 1992.

A Comprehensive Approach for the Evaluation of Chemicals for Asphaltene Deposit Removal

L. Barberis Canonico, A. Del Bianco, G. Piro, and F. Stroppa

ENIRICERCHE SPA, MILAN, ITALY

C. Carniani and E. I. Mazzolini

AGIP SPA, MILAN, ITALY

1. INTRODUCTION

Asphaltene deposition during petroleum extraction may reduce the oil production up to the point where remedial work is required to remove the deposit and restore production. In the case of asphaltene deposits within the formation, the injection of aromatic solvents, in some case containing chemical additives, has been employed to recover well productivity. The most widely used solvents range from toluene to light petroleum distillates /1,2/ while the additives include different classes of products such as aliphatic amines, alkyl benzene sulfonic acids, phenols, polymers of different nature,etc. /3-5/.The correct choice of the solvent and/or additive, by means of a preliminary screening of the solvent activity, is crucial to assure the success of a treatment. The literature reports a number of studies aimed at evaluating the effectiveness of products for removing asphaltene deposits /2,6-7/. Most of these describe procedures for generating comparative asphaltene solvency data under standard conditions. However, significant differences in performance may be observed upon changing the reference conditions (eg. sample to solvent ratio, contact time and temperature). Dissolution experiments on isolated bulk asphaltene samples ignore the existence of an adsorbed asphaltene layer on the formation rock surface and therefore fail to consider the effect of the asphaltene-mineral interaction. The adsorbed asphaltene layers may affect oil production either indirectly, by changing the oil/water relative permeability, or directly, by favouring the asphaltene aggregation/flocculation phenomena within the oil medium /8/. In our opinion, underestimation of the specific asphaltene/formation rock interactions within the problem formation may be detrimental to the achievement of a satisfactory remedial treatment. Relatively little insight into asphaltene-rock interaction has appeared in the literature, and even less on the efficacy of solvents and additives in removing the adsorbed asphaltene; consequently, in the course of optimizing the remedial treatment for a specific field problem, we have developed a general procedure for screening solvents which takes into account characteristics of both bulk and adsorbed asphaltene. In this paper we define standard procedures for evaluating solvent/additive performance in terms of maximum solvent capacity and rate of dissolution at different testing temperatures. In addition, the problem of adsorbed asphaltene onto reservoir rock has been addressed in a series of core floods, and the efficacy of removal with different solvents and additives has been studied.

2. EXPERIMENTAL

2.1 Experimental Materials

Asphaltene Samples. This study has employed asphaltenes recovered from a well in northern Italy during treatment of the tubing and formation with toluene. The solvent suspension containing both dissolved and undissolved asphaltic material was evaporated and then Soxhlet-extracted with n-heptane. The n-heptane/insoluble fraction (HI) was further extracted with THF (THFS) leaving behind the inorganic part of the deposit. The HI-THFS fraction was isolated by evaporation, dried overnight at 80 °C under vacuum and utilized as the reference sample in this study. Core-flood experiments also were carried out with asphaltenes isolated from the crude oil by precipitation with n-heptane. The chemical properties of these samples are the following:

- HI-THFS residue: H/C = 0.778; ^{13}CNMR ar.factor = 0.75; Mw = 1600 amu
- C7 asphaltene: H/C = 0.815; ^{13}CNMR ar.factor = 0.69; Mw = 1900 amu

Solvents. Reagent grade toluene, xylene, n-propyl-benzene, tetraline and 1-methyl-naphthalene and three commercial solvents (S1-3) were employed as received (Table1).

Commercial Additives. Additive A, based on alkyl benzene sulfonic acid, and additives B and C, based on complex polymers, were used as received from the manufacturer.

Core material. Porous media studies were performed on powered reservoir dolomite sand-packs (average particle size = 60 mμ, BET surface area = 10 m²/g).

2.2 Asphaltene Dissolution Studies

Bulk Asphaltene Dissolution. Dissolution experiments were carried out on both powdered asphaltene and compressed asphaltene pellets. The maximum capacity of solvents to dissolve asphaltene was determined at room temperature by diluting 100 mg of the powdered asphaltenes with increasing amounts of solvent and determining the dissolved asphaltene at equilibrium (i.e., after sonication and 12 hours stirring). Asphaltene concentrations were measured on filtered solutions (0.5 μm teflon filter) by a UV-VIS spectroscopic method using calibration curves. Owing to the complex nature of asphaltenes, (large distribution of molecular masses) the UV-VIS spectrum may change not only as a function of the amount but also, to some degree, as a function of the quality of the dissolved material (different ε). For this reason, asphaltene concentrations were obtained by averaging the concentrations calculated at three wavelengths (400, 600 and 800 nm).The experiments aimed to assess the rate of dissolution of solvents (kinetic curves) were carried out on standard asphaltene pellets (cylinders of 13 mm diameter x 0.7 mm, obtained with 100 mg of sample pressed at 10.000 Kg/cm^2).The pellets were immersed in a fixed amount of solvent (1L) and maintained at constant temperature under mild agitation in a glass autoclave. The supernatant was analysed periodically by removing samples of a few mL.The experiments with light solvents at high temperature (up to 150°C) were carried out under nitrogen pressure (maximum 0.5 MPa) to maintain the solvent in the liquid state. Both the solubility and the kinetic curves were drawn by fitting the experimental data (7 points minimum) by an interpolation routine.

Adsorbed Asphaltene/Core-Flood Experiments. Flooding experiments were conducted by means of the apparatus shown in Figure 1.

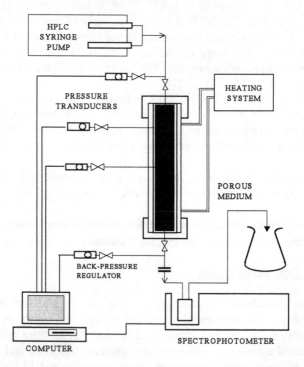

Fig. 1   Porous medium apparatus

The dolomia sand-packs were confined in stainless-steel core holders (1,27 cm diameter, 10 cm length) equipped with pressure transducers allowing measurement of the pressure-drop at different positions along the packing. The sand-packs showed permeabilities ranging from 2000 to 3000 mD. After passing through the porous medium the experimental fluids were analyzed on-line with an $\Lambda 2$ Perkin-Elmer UV cell (quartz, 1 mm optical path length). In this way asphaltene concentrations were determined spectroscopically as described above. The experimental parameters were controlled and data recorded and analyzed for the quantity of both adsorbed and removed asphaltenes by computer. High temperature (150°C) experiments were performed by circulating hot silicon oil through a jacketed core holder external to the sand-pack. The presence of a single solvent phase within the porous medium was assured with a back-pressure regulator placed between the core holder outlet and the UV detector. The core-flood experiments were carried out with the following procedure:

1. Adsorption stage: a 1000 ppm asphaltene solution in a solvent (toluene or trichloroethylene) was flushed at 20 ml/h through the dolomia sand-pack until the effluent asphaltene concentration, C^*, reached the in-let value, $C°$ (Figure 2). This saturation state was usually attained after 20 pore volumes (1 pore volume = 5-6 ml).

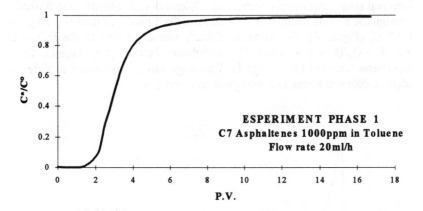

Fig. 2   Plot of asphaltene concentration (deposition flooding)

2. Washing stage: pure trichloroethylene or toluene solvent was flushed at 40 ml/h through the porous medium in order to displace the asphaltene solution from stage 1 and to remove the asphaltene material weakly adsorbed onto the powdered rock. This stage was terminated only after the asphaltene

concentration in the effluent dropped below the detection limit (usually after 10 pore volume) (Figure 3).

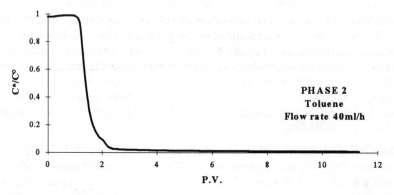

Fig. 3   Plot of asphaltene concentration (removal flooding with solvent)

3. Removal stage: commercial solvent or additive-doped toluene was flushed through the porous medium at 40 ml/h for a fixed period of time (10-20 p.v.= 1.5-3 h) (Figure 4). The removal efficacy was calculated in the following way: $E = Q_3/Q_1 \times 100$, where, Q_1= asphaltene adsorbed (mg, stage 1), Q_2= asphaltene removed (mg, stage 3). This stage can be performed at different additive concentrations and sand-pack temperatures.

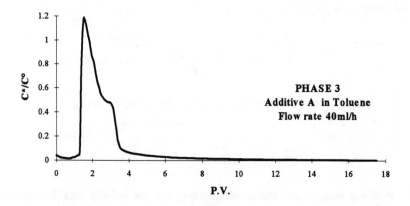

Fig. 4   Plot of asphaltene concentration (removal flooding with additives)

3. RESULTS AND DISCUSSION

3.1 Solvent Comparison

Bulk Asphaltene Dissolution. The evaluation of solvents was conducted with both pure model solvents and commercially available products. The solubility curves of the aromatic hydrocarbons (Figure 5) show a very distinctive behaviour. Alkylbenzenes, whose peculiar solvent capacity may be observed especially at high

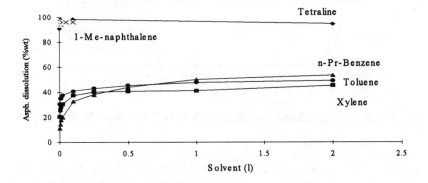

Fig. 5 Solubility curves for model compounds

sample to solvent ratios, dissolve at most 50% of the sample, while higher-condensed aromatic hydrocarbons (HCAH) such as tetraline and 1-methyl-naphthalene gave more than 90% dissolution. Examination of the rate of asphaltene dissolution confirm further the much higher efficency of the latter solvents over alkylbenzenes (Figure 6).

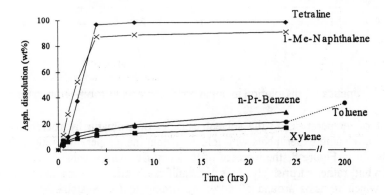

Fig. 6 Kinetics of dissolution in model compounds at room temperature

In these experiments, the dissolving kinetics of toluene are so slow that even after 200 hours, it does not achieve its equilibrium solvent capacity. The results of these experiments show clearly that for asphaltic material, relatively small variation of the solubility parameter (δ) can significantly influence the performances of the solvent. The same tests carried out with three commercial solvents show that their behaviour may be related to the products composition (Table 1 and Figure 7). Once again the solvent power improves as the concentration of HCAH in the mixture increases.

TAB.1 Composition by categories of commercial solvents

Composition\Solvent	S1	S2	S3
Saturate	-	1.1	-
Alkylbenzenes	11.4	77.5	81.4
Naphthenobenzenes	7.3	10.9	16.5
Naphthalenes	81.3	10.5	2.1

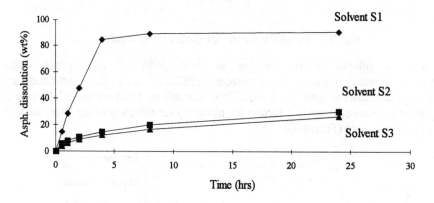

Fig. 7   Kinetics of dissolution in commercial solvents at room temperature

The effect of temperature on the solvent capacity of aromatic hydrocarbons was examined performing experiments with toluene within the temperature range 20-150 °C. As shown in Figure 8, the increase of temperature sharply enhance the rate of dissolution but, rather surprisingly, does not significantly affect the maximum solvent capacity, which remains around 50-60%. According to the Scatchard-Hildebrand equation /9/, asphaltene-solvent mutual solubility is governed by the difference in the respective solubility parameters (Δδ). It is evident that for the asphaltene-toluene system examined this term remains almost constant, at least in the range 20-150 °C.

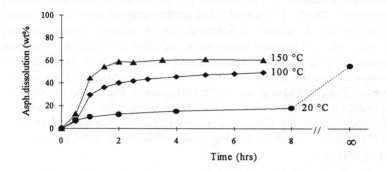

Fig. 8   Kinetics of dissolution in toluene at different temperatures

Adsorbed Asphaltene Removal. Based on the data previously discussed, solvents were selected and investigated in order to evaluate the adsorbed asphaltene removal efficacy: toluene, as representative of the alkyl benzene class of solvents and 1-methyl-naphthalene as representative of HCAH. The solvents asphaltene removal efficacy was determined with respect to asphaltenes from well deposits (i.e., HI-THFS residue) at room temperature and 150°C. The experimental results reported in Table 2 show that 1-methyl-naphthalene always exhibits an higher removal efficacy with respect to toluene.

Tab.2 Asphaltene removal at different temperatures

Solvent	Asphaltenes Adsorption (mg)	Removal % at room T	Removal % at 150°C
toluene	15.8	1	9.7
1-methyl-naphthalene	15.7	25.6	58

Although this finding agrees with the results obtained for bulk asphaltenes, the level of removal that these solvents can attain when the same asphaltic material is in the rock adsorbed form is quite different. In particular, here 1-methyl-naphthalene achieves only 58% removal of the asphaltene. The observed lower performance of these solvents is evidently attributable to the different interaction forces operative in the case of bulk and adsorbed asphaltenes.

3.2 Additives

Bulk Asphaltenes Dissolution. The performance of additives was evaluated (2% by weight in toluene) as described previously. The results obtained show that additives of different nature can have very different effects on toluene solvent capacity (Figures 9, 10, 11). At ambient temperature, additive B enhanced   toluene performance significantly, additive C increased the solvent capacity of toluene but did not influence the kinetics of dissolution, additive A was practically ineffectual. As for toluene alone, increasing of the temperature affected the kinetics of asphaltene dissolution but did not alter the maximum solvent capacity of the system. These results suggest that the dissolution and removal of asphaltene deposits is favoured by the action of products bearing polar (additives B and C) but not H^+ donors group (additive A).

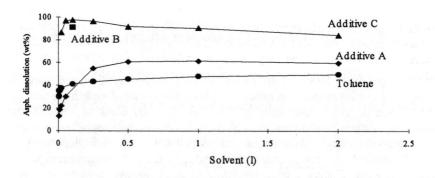

Fig. 9   Solubility curves with Additives (2%wt in Toluene)

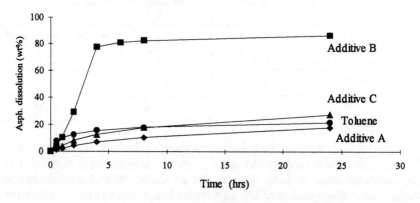

Fig. 10   Kinetics of dissolution with Additives
(2%wt in Toluene at room Temperature)

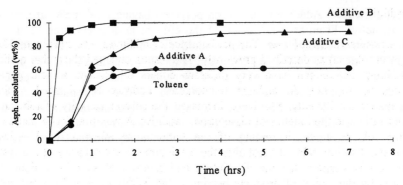

Fig. 11 Kinetics of asphaltene dissolution with additives (2%wt in Toluene at 150°C)

Adsorbed Asphaltene Removal. The removal efficiency of some commercial additives in toluene were also examined in core-flooding experiments. The effect of additive concentrations in the range 0.1 - 4 wt.% was first evaluated at room temperature working with n-heptane precipitated asphaltenes (Table 3).

Tab.3 Asphaltene removal at different additive concentrations (room temperature)

Additive	Additive concentration (%)	Asphaltene adsorption (mg)	Asphaltene removal (%)
A	2.0	11.4	97.2
	0.2	13.4	88.5
	0.1	12.7	81.2
B	2.0	12.9	89.4
	0.5	11.9	87.7
	0.1	11.9	70.7
C	4.0	13.9	82.6
	2.0	12.6	80.2
	0.5	13.8	59.6

The results reported in Table 3 show that all three additives studied gave better results with respect pure toluene. The additive based on alkyl benzene sulfonic acid (additive A) is the most active, reaching a removal efficacy of 80% at a concentration of only 0.1%. Additive B, polymers bearing polar groups, also shows good performance. Additive C, based on polymers having weak polar groups, was the least active. From an examination of the effluent profiles (Figures 12, 13, 14) possible removal mechanisms of these additives can be inferred. For additive A (Figure 12) we observe that on increasing additive concentration (from 0.1% to 0.2%) almost the same amount of removed asphaltene is achieved within a period of time that is roughly double. In other words, the additive concentration does not affect the quantity of asphaltenes removed, rather it affects the removal action time. This fact suggests that competitive interaction of additive with the rock active sites might be the mechanism by which the asphaltenes are desorbed from the rock surface.

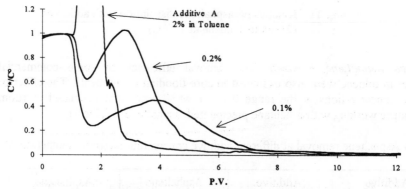

Fig. 12   Outlet of asphaltenes concentration in removal flooding with additive A dissolved in toluene

Considering additive C (Figure 13) we point out that the amount of asphaltene removed increases with additive concentration within the same period of time.

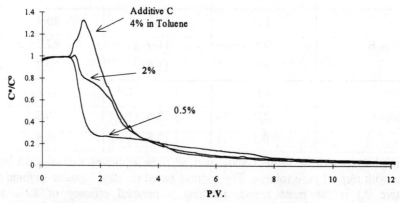

Fig. 13   Outlet of asphaltenes concentration in removal flooding with additive C dissolved in toluene

Tentatively, this behaviour suggests that additive C might function as a "solvent enhancer". In other words this additive, by increasing the solvation energy of the asphaltene in solution, shifts the thermodynamics in favor of desorption.

In the case of additive B (Figure 14), whose shape is intermediate between those of additives A and C, we assert that the removal mechanism is based both on the competitive interaction with the rock surface active sites and on enhancement of the solvent power.

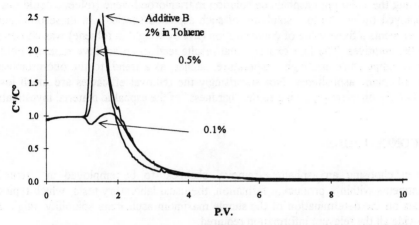

Fig. 14 Outlet of asphaltenes concentration in removal flooding with additive B dissolved in toluene

High temperature experiments have also been performed at additive concentrations A, 0.1% , B, 0.1% , C, 0.5% (w/w). The results obtained are reported in Table 4.

Tab. 4 Asphaltene removal with additives at 150°C

Additive	Additive Concentration	Temperature (°C)	Asphaltenes Adsorbed (mg)	Removal (%)
Toluene	-	room T	11.9	11
Toluene	-	150	11.9	46.6
Additive A	0.1% in Tol	150	11.1	86.6
Additive B	0.1% in Tol	150	10.1	99.9
Additive C	0.5% in Tol	150	10.8	78.8

At 150°C the additive ranking was: B > A > C. At high temperature the performances of both additive B and C improved, in agreement with both solubilization and desorption process thermodynamics. The lower performance of additive A might be attributed, tentatively, to the more severe depletion of the additive acid group, owing to a faster high-temperature acid-base reaction between the additive and dolomia. We have also studied asphaltene removal by additives in core flood experiments carried out both at room temperature and 150°C with hard asphaltene deposit. These asphaltenes were introduced into the porous medium by flowing the 1000 ppm asphaltene solution in trichloroethylene (toluene could not be employed owing the low solubility of such asphaltic species). In these preliminary experiments a fixed value of concentration (2% by weight in toluene) was chosen for all the additives. The first experimental results lead to an additive ranking, both at room temperature and high temperature, similar, as a trend, to the one established for n-heptane asphaltenes. Not surprisingly the removal efficacies are much lower (30 - 60% on average) owing to the "hardness" of the asphaltic material investigated.

4. CONCLUSIONS

When choosing an asphaltene-removal chemical to be employed in remedial treatments within a producing formation, the usual laboratory tests, which typically focus on the determination of the simple maximum asphaltene solubility, might not provide all the relevant information required.

The experimental results obtained in this work have shown how the selection of an effective solvent system is rather complex, owing to the dual nature of the asphaltic material (i.e., bulk and adsorbed asphaltenes) and the peculiarity of the processes occuring in terms of both dissolution kinetics and thermodynamics. For a particular asphaltene-damaged formation of interest, the tests elaborated here allow one to establish a ranking amongst different asphaltene-removal chemical systems with regard to:

- bulk and adsorbed asphaltene dissolving power
- rate of removal (contact time)
- temperature (of interest)

From the experimental data obtained through the reported tests two different situation may come out:

1. Bulk and adsorbed asphaltenes present the same solvent removal ranking.

2. Bulk and adsorbed asphaltenes do not present the same solvent removal ranking.

The first case allows the straightforward choice of top-rank chemical to be employed in remedial treatments. The second case requires more reasoning in making the right choice of the best suited removal chemical: one may choose the remedial chemical that represents the best compromise amongst bulk and adsorbed asphaltenes or, knowing the well history, one may choose the chemical best suited either for bulk or adsorbed asphaltenes. The validation of the choice would be provided by the post-treatment results.

5. ACKNOWLEDGMENT

The authors would like to thank Dr.T.P.Lockhart from ENIRICERCHE SpA for stimulating discussions and the help he provided during the preparation of this paper.

6. REFERENCES

1. C.W. Benson, R.A. Simcox, I.C. Huldal, Fourth Symposium on "Chemicals in the Oil Industry: Dev. & Appl.", Ed P.H.Ogden, 215, **1991**

2. G. P. Dayvault and D.E. Patterson, SPE 18816, SPE Reg. Meeting, Bakersfield, CA, April 5-7, **1989**

3. M.L. Samuelson, SPE 23816, SPE Int. Symp. on Formation Damage, Lafayette, Louisiana, Feb. 26-27, **1992**

4. G. Gonzales, A. Middea, Colloids and Surfaces, 42, 207 (**1991**)

5. G. Broaddus, J. of Petr. Techn. , June **1988**, 685

6. M.G. Trbovich, G.E. King, SPE 21038, SPE Int. Symp. on Oilfield Chem., Anaheim, Feb. 20-22, **1991**

7. M.E. Newberry, K.M. Barker, SPE Prod. Op. Symp., Oklahoma City, March 10-12, **1985**

8. G.Gonzales and A.M.T.Luovisse, SPE 21039, SPE Int.Symp. on Oilfield Chemistry, Anaheim, California, Feb. 20-22, **1991**

9. S.I. Anderson, K.S. Birdi, Fuel Sci. & Techn. Int'l., 8(6), 593, **1990**

Coprecipitation Routes to Absorbents for Low-Temperature Gas Desulphurisation

T. Baird, K. C. Campbell, P. J. Holliman, R. Hoyle, and D. Stirling

CHEMISTRY DEPARTMENT, UNIVERSITY OF GLASGOW, GLASGOW, G12 8QQ, UI

B. P. Williams

ICI KATALCO, BILLINGHAM, CLEVELAND, UK

Abstract

The hydrogen sulphide absorption capacity of Co/Zn and Co/Zn/Al oxides was determined using a continuous flow absorption rig. The oxides were prepared by decomposition of the mixed metal basic carbonates formed by a coprecipitation route. The H_2S absorption capacity increased with increase in cobalt concentration in the two component systems. High uptake was associated with the normal spinel Co_3O_4, a surface concentration of cobalt, a high surface area and the formation of microcrystalline membraneous sheets containing cobalt, zinc and sulphur. The Al_2O_3 was thought to have a deleterious effect on the H_2S uptake either by partial substitution of aluminium in the Co_3O_4 spinel or by partial substitution of zinc or cobalt in the Al_2O_3 structure. The three component systems formed hydrotalcite precursors which decomposed to give high surface area interdispersed oxides, but these were poor absorbents for H_2S.

Introduction

Sulphur compounds are found as natural contaminants in many hydrocarbon feedstocks used in metal-catalysed industrial processes. Many transition-metal catalysts, such as the supported nickel catalyst used in steam reforming, are poisoned by sulphur compounds which behave as Lewis-type bases by donating electrons into the unfilled d-orbitals of the metal.[1] Sulphur compounds in feedstocks can also limit plant lifetime by causing pipeline corrosion and can have a detrimental effect on the environment by contributing to acid rain when emitted into the atmosphere.[2] There is therefore a clear need for efficient desulphurisation of feedstocks and clean-up of refinery effluent.

Industrially, sulphur-containing organic compounds can be efficiently converted to H_2S using a $CoO/MoO_3/Al_2O_3$ catalyst at 643 K, 40 bar and the H_2S can then be absorbed in a bed of zinc oxide at 623 K.[3] Absorption of H_2S by zinc oxide is stoichiometric at this temperature but falls off rapidly as the temperature is lowered. The development of a high surface area zinc oxide [4] has enabled the H_2S absorption to be carried out at temperatures less than 473 K which has considerable economic advantages, but uptake is lowered. Previous work by this group[5] has shown that the H_2S uptake is dependent both on the transition metal ion used and the method by which the absorbent is prepared. Precipitation routes have been found to give precursors that decompose to give the highest surface area metal oxides.

There are three main precipitation techniques used for mixed metal salts, viz. sequential precipitation of ions, precipitation at constant pH and high supersaturation, and coprecipitation at constant pH and low supersaturation.[6]

In the sequential method the alkali is added to transition metal nitrate salts and metal hydroxides are precipitated out sequentially as the pH rises. One disadvantage with this method is that the components tend to precipitate out in more than one phase. Furthermore, even if a single phase is formed it is frequently the basic nitrate rather than the carbonate that is precipitated. Thus, Petrov *et al.*[7] found that addition of alkali to zinc and cobalt nitrates resulted in the formation of a basic cobalt/zinc hydroxynitrate species $[Zn_{1.66}Co_{3.34}(OH)_{8.82}(NO_3)_{1.26}(H_2O)_{2.23}]$. Formation of hydroxycarbonates is generally preferable to the formation of nitrates and hydroxynitrates since the decomposition of the latter is accompanied by the evolution of toxic nitrogen oxides. Hydroxycarbonates, on the other hand, are low cost materials that decompose at low temperatures without the evolution of toxic gases to give high surface area oxides.

Precipitation at constant pH and high supersaturation involves rapid addition of metal nitrate solutions in high concentrations to a base. Under these conditions of high supersaturation the rate of nucleation is much higher than that of crystal growth, and this results in rapid formation of a large number of small particles whose composition may not be uniform. The high number of crystallisation nuclei generally means that the precipitate will be amorphous.[8]

The third technique, coprecipitation at low supersaturation and constant pH, has been used extensively in the preparation of one-, two- and three-component basic carbonate precursors.[6,9] In this method, mixed metal ions are precipitated out as their hydroxycarbonates by simultaneous addition of an alkali carbonate. Coprecipitation was found to give the best interdispersed high surface area mixed oxides on calcination.[9] Coprecipitation techniques have therefore been adopted in this work. The reason for the success of the coprecipitation route is that it enables the metals to be incorporated in the same crystalline structure, the homogeneity of which can be maintained by appropriate selection of calcination conditions.

The composition of the precursor and ultimately the mixed metal oxide is dependent on many factors such as the pH and temperature at which the precipitation reaction is carried out, the concentrations and ratios of the metal salts and the ageing time of the precipitate in the mother liquor.

The pH at which coprecipitation is carried out is critical in determining the precipitated phase or phases. For the preparation of hydroxycarbonates containing two or more cations, it is necessary to coprecipitate at a pH higher than or equal to that at which the more-soluble hydroxide precipitates. Thus, in the preparation of Co/Zn and Co/Zn/Al basic carbonates discussed in this paper this generally means working in the pH range 7-10 [6], but even within this range the precipitated phases can vary. At pH greater than 12, the dissolution of aluminium and formation of zincate $[ZnO_2^{2-}]$ occur, resulting in a different precipitated phase.

The composition of the precursor is also dependent on the ratios of the metal ions. Porta *et al.* [9] prepared Co/Cu hydroxycarbonates by coprecipitation at pH 8 using sodium hydrogen carbonate as the base. Precursors with a Co/Cu ratio of < 33/67 formed a cobalt-containing malachite whereas copper-containing spherocobaltite and spherocobaltite phases were formed for Co/Cu ratios of 85/15 and 100/0 respectively. The cation ratio is particularly important in the preparation of mixed metal hydroxycarbonates with the hydrotalcite structure. Natural hydrotalcites have the general formula $[M(II)_6M(III)_2(OH)_{16}CO_3.4H_2O]$ and are comprised of positively charged sheets of metal hydroxides containing two metals in different oxidation states. [10] The carbonate anions and water molecules are located between these metal hydroxide layers (Figure 1). Most of the bivalent metals from Mg^{2+} to Mn^{2+} and all the trivalent ions except V^{3+} and Ti^{3+} with atomic radii 0.05 to 0.08 nm form hydrotalcites. The hydrotalcites are obtained pure for an M(III)/M(III)+M(II) ratio of 0.2 to 0.33. [6] Other basic carbonates, hydroxide and oxide phases are formed outwith this range.

Finally, decomposition of the precursors should be carried out at the minimum temperature required to form the oxide if a high surface area is the main prerequisite, as is the case in this work. Higher temperatures result in increased interactions between mixed metal oxides and can lead to the formation of ordered spinels.

A preliminary study of mixed cobalt/zinc oxides prepared by a coprecipitation route[5] showed that doping zinc oxide with cobalt oxide could improve its H_2S absorption capacity. These investigations have now been extended by preparing a range of Co/Zn and Co/Zn/Al mixed oxides from their hydroxycarbonate precursors and testing them for their H_2S absorption capacity. The preparation, characterisation and testing of these oxides are discussed in this paper.

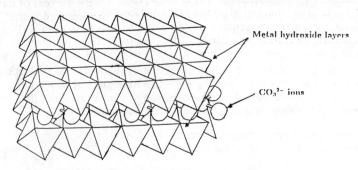

Figure 1:- Schematic Diagram of the Hydrotalcite Structure.

Experimental

1. Absorbent preparation.

1 mol dm^{-3} solutions containing cobalt/zinc or 0.5 mol dm^{-3} of cobalt/zinc/aluminium nitrates in the ratios given in Table 1 were pumped simultaneously with a 2 mol dm^{-3} carbonate solution into a mixing vessel at 353 K, the flow of carbonate being adjusted so that precipitation occurred at a constant pH of 7.0. The precipitate was aged in the mother liquor for 30 minutes, filtered and washed in deionised water to remove any impurities. The precipitates were all dried at 383 K for 16 hours. The Co/Zn precursors were calcined in air for 16 hours at 623 K. The Co/Zn/Al precursors were heated in flowing air at 5 K min^{-1} from 293 K to 723 K and then held at this temperature for 4 hours prior to cooling. The coprecipitated precursors will be referred to as Co/Zn or Co/Zn/Al with a suffix indicating the nominal metal ratios. The suffix "cal" will be added for the calcined samples.

2. Absorbent characterisation

a) Metal loadings for the precursors and oxides were determined by atomic absorption using a Perkin-Elmer 370A spectrometer.
b) X-ray Powder Diffraction (XRD) measurements were made with a Philips diffractometer and Ni-filtered Co $K\alpha$ radiation ($\lambda = 0.1790$ nm).
c) Samples were examined on a JEOL 1200 EX transmission electron microscope. Specimens were prepared by suspending them in water and then mounting them on carbon-filmed copper grids.
d) BET surface areas (N_2, 77 K) were obtained from five-point adsorption isotherms using a Flowsorb 2300 Micromeritics instrument.

3. Absorbent testing.

The H_2S absorption capacity of the calcined precursors (sieved to particle sizes between 500 and 1000 um) was determined using the flow system detailed in Figure 2. A gas mixture containing 2% H_2S in N_2 was delivered at a space velocity of 700 hr^- from mass flow controllers to a 5 cm^3 bed of the oxide(s) in a Pyrex glass reactor [1]. The reactor exit gas was bubbled through a trap containing alkaline lead acetate. The time interval from the initial contact of the H_2S/N_2 carrier with the absorbent bed to "breakthrough" of H_2S at the bed exit was measured. Breakthrough was determined by the detection of 2-3 ppm H_2S in the exit stream measured by the precipitation of lead sulphide in the alkaline lead acetate. Measurement of the concentration of H_2S in the feed stream and the growth in concentration of H_2S in the reactor exit after breakthrough were determined by on-line G.C. analysis.

Results

1. Absorbent precursors.

The atomic absorption results and phases detected for each metal ratio are presented in Table 1. The metal loadings measured by atomic absorption were

Table 1: Absorbent Phases and Compositions.

Nominal Co/Zn/Al	Actual Co/Zn/Al	Precursor Phase-type[‡]	Oxide Phase-type[‡]
0/100/0	0/100/0	Hz[*]	ZnO
10/90/0	9.7/90.3/0	Hz	ZnO + Co_3O_4 (trace)
20/80/0	23.9/76.1/0	Hz + Sph[†]	ZnO + Co_3O_4
30/70/0	28.9/71.1/0	Hz + Sph	ZnO + Co_3O_4
40/60/0	41.9/58.1/0	Hz + Sph	ZnO + Co_3O_4
50/50/0	43.8/56.2/0	Sph	ZnO + Co_3O_4
90/10/0	90/10/0	Sph	Co_3O_4
100/0/0	100/0/0	CoX	Co_3O_4
15/60/25	16.0/62.3/17.7	Htc[°]	ZnO
37.5/37.5/25	37.4/37.5/20.0	Htc	Co_3O_4
60/15/25	68.0/16.2/20.4	Htc	Co_3O_4

[‡] Where a mixed-metal phase is present it is assumed that all metals are dissolved in all the phases (at least to some extent).

[*] Hz = Hydrozincite-type phase.

[†] Sph = Spherocobaltite-type phase.

[°] Htc = Hydrotalcite.

X = $(CO_3)_{0.5}(OH)_{1.0}0.11H_2O$.

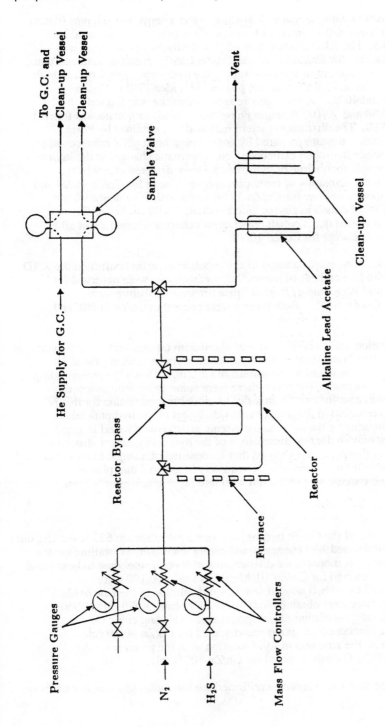

Figure 2: The H_2S Testing Line.

in good agreement with the nominal loadings. The interplanar spacings (d/nm) derived from X-ray diffraction data for each of the precursors are listed in Tables 2 and 3. The XRD studies showed that the hydrozincite phase was formed exclusively for Co/Zn 0/100 and Co/Zn 10/90. It can be assumed that all the cobalt was present in solid solution in the hydrozincite in Co/Zn 10/90. Both hydrozincite and spherocobaltite phases were identified in Co/Zn 20/80, 30/70 and 40/60. A monophasic spherocobaltite was formed for Co/Zn 50/50 and 90/10. A single phase basic cobalt carbonate was formed for Co/Zn 100/0. The diffraction pattern matched that obtained by Porta [11] for a basic cobalt carbonate prepared by coprecipitation at pH 8 from cobalt nitrate and sodium hydrogen carbonate. No systematic changes in the lattice parameters for spherocobaltite or hydrozincite were detected in any of the precursors, but the formation of monophasic hydrozincite for Co/Zn 10/90 and monophasic spherocobaltite for Co/Zn 50/50 and 90/10, is evidence for the partial solubility of cobalt in zinc or zinc in cobalt. XPS studies of the precursors [12] revealed that a cobalt enrichment occurs at the surface of all these compounds except for Co/Zn 10/90.

TEM studies of selected samples of the cobalt/zinc series confirmed the XRD results. Needle-like crystals of monoclinic hydrozincite were observed for Co/Zn 0/100 and 10/90 and diffraction lines for spherocobaltite were identified for Co/Zn 50/50. Both phases were present in Co/Zn 20/80, and 40/60.

Coprecipitation using cobalt, zinc and aluminium nitrates and sodium hydrogen carbonate resulted in the synthesis of hydrotalcite type structures (Table 3). Additional lines were found in the pattern of Co/Zn/Al 15/60/25 corresponding to that of zinc hydroxide. However, there were some systematic absences in the diffraction data for zinc hydroxide. A possible explanation for this is that there are areas within the metal hydroxide layers of the hydrotalcite structure approximately the same as the atomic arrangement found in zinc hydroxide, but only in the two dimensions of the hydrotalcite layer structure. TEM studies of this precursor showed that it consisted of a single hydrotalcite phase. The particles consisted of approximately hexagonal thin platelets with an average particle size of 5.0 nm with interlayer spacings of 0.7 nm.

2. Oxides.

Decomposition of the Co/Zn hydroxycarbonate precursors at 623 K for 16 hours in air gave well defined XRD patterns, indicating that highly crystalline oxides had been formed. A monophasic diffraction pattern corresponding to hexagonal zinc oxide was detected for Co/Zn 10/90 calc as well as 0/100 calc, indicating that all the cobalt was present in solid solution in the zinc oxide. Biphasic precursors were obtained for Co/Zn 20/80 calc, 30/70 calc, 40/60 calc and 50/50 calc, consisting of hexagonal zinc oxide and cubic Co_3O_4. A single phase zincian cobalt oxide was detected for Co/Zn 90/10 calc indicating that all the zinc was in solid solution as in the precursor. A single phase of Co_3O_4 was detected for Co/Zn 100/0 calc.

The electron diffraction results confirmed the formation of a single zinc oxide

Table 2: D-Spacings of the CoZn Precursors from XRD (Values are in Angstroms)

Zn(CO₃)₂(OH)₆†	h k l	CoCO₃*	h k l	CoX‡	100/0	90/10	20/80	30/70	40/60	50/50	90/10	100/0
6.77 (100)	200				6.77 (100)	6.77 (100)	6.77 (60)	6.77 (100)	6.86 (100)			
5.37 (10)	001			5.06 (70)		5.37 (8)						5.08 (84)
				4.44 (5)								4.49 (40)
3.99 (20)	-111				4.01 (17)	4.02 (20)	3.98 (21)	4.04 (28)	4.01 (26)	3.96 (20)		
3.66 (40)	-310				3.70 (15)	3.70 (15)		3.67 (12)				
		3.55 (40)	012				3.56 (39)	3.53 (19)	3.58 (51)	3.56 (45)	3.56 (40)	
												3.42 (44)
3.14 (50)	020				3.16 (40)	3.17 (43)	3.15 (30)	3.17 (38)	3.17 (65)	3.09 (32)		
3.00 (10)	-401											
2.92 (20)	311				2.94 (10)	2.96 (12)						
2.85 (30)	220				2.87 (22)	2.87 (31)	2.87 (16)	2.87 (22)				
2.74 (10)	401	2.74 (100)	104				2.74 (100)		2.75 (100)	2.76 (100)	2.74 (100)	
2.72 (60)	021				2.72 (61)	2.72 (74)		2.72 (75)				
2.69 (20)	002	2.65 (100)		2.65 (100)	2.67 (7)	2.68 (31)						2.64 (100)
2.58 (10)	-202	2.53 (70)		2.53 (70)	2.57 (12)	2.58 (20)	2.56 (18)					2.53 (65)
2.48 (70)	510				2.49 (40)	2.49 (41)		2.50 (42)	2.50 (42)			
2.30 (20)	-420	2.33 (20)	110	2.295 (70)	2.31 (9)	2.32 (20)						2.28 (81)
										2.26 (11)		
2.21 (10)	312				2.21 (8)	2.20 (12)						
		2.11 (20)	113				2.10 (27)		2.12 (16)	2.13 (20)	2.12 (20)	
1.92 (30)	222	1.95 (20)	202	1.92 (40)	1.92 (15)	1.92 (17)	1.92 (24)		1.95 (26)	1.95 (19)	1.95 (29)	1.91 (42)
1.90 (30)												
		1.78 (10)	024							1.78 (10)		
1.69 (40)	800	1.70 (30)	116	1.69 (10)	1.69 (12)	1.70 (12)	1.71 (30)	1.70 (23)	1.72 (40)	1.71 (40)	1.71 (40)	1.70 (33)
1.57 (20)	-622						1.58 (20)					
1.56 (10)	023			1.55 (20)			1.55 (26)		1.56 (40)			1.55 (26)

† JCPDS 19–1458.   * JCPDS 11–692.   X = (CO₃)₀.₅(OH)₁.₀0.11H₂O.

‡ P. Porta et al, J. Chem. Soc., Faraday Trans, 1992, **88**(3), 311.

Intensities (I/I_0) are shown in brackets.

Table 3: D-Spacings of the CoZnAl Precursors from XRD (Values are in Angstroms)

Mg_6Al_2X	h k l	$Co_{1.2}Zn_{4.8}Al_2X$	$Co_3Zn_3Al_2X$	$Co_{4.8}Zn_{1.2}Al_2X$
7.69 (100)	003	7.3973 (100)	7.4507 (100)	7.5597 (100)
		4.4897 (10)	4.4705 (10)	
3.88 (70)	006	3.7392 (23)	3.7659 (26)	3.7930 (30)
		3.3181 (8)	3.2976 (9)	
		2.6234 (10)	2.6297 (13)	
2.58 (20)	012	2.5679 (36)	2.5739 (42)	2.5800 (32)
		2.4586 (3)		
2.30 (20)	015	2.2815 (20)	2.3998 (4)	2.2956 (18)
			2.0908 (20)	
1.96 (20)	018	1.9353 (12)	1.9353 (13)	1.9516 (8)
1.85 (10)	0012			
1.75 (10)	1010			1.7379 (2)
1.65 (10)	0111			
				1.6007 (2)
1.53 (20)	110	1.5321 (8)	1.5321 (11)	1.5358 (11)
1.50 (20)	113	1.5022 (6)	1.5048 (9)	1.5066 (12)

$X = (OH)_{16}CO_3.4H_2O$.

phase for Co/Zn 0/100 calc and a single Co_3O_4 phase for Co/Zn 100/0 calc.
Co_3O_4 was also found to be the only phase present in Co/Zn 50/50 calc and ZnO
then appears with increasing intensity in the diffraction patterns for the more
zinc-rich samples ie Co/Zn 40/60 calc and 20/80 calc. A faint signal for either
Co_3O_4 was also detected in Co/Zn 10/90 calc. It would appear that the
solubility of Co in ZnO is less than that of Zn in Co_3O_4. Notwithstanding
this,the formation of single phases of zinc oxide or Co_3O_4 in the presence of
10% cobaltand zinc respectively indicates that the two phases found at
intermediate compositions may contain appreciable amounts of zinc or cobalt.
This has yet to be confirmed. However, XPS data [12] have shown that the
normal spinel $ZnCo_2O_4$ is present at the surface at Co/Zn ratios of
$\leq$ 30/70. Although a cobalt enrichment was detected at the surface in these
oxides as in the precursors, XRD studies clearly showed that the $ZnCo_2O_4$
spinel phase did not extend into the bulk.

The transmission electron micrograph for Co/Zn 20/80 calc is shown in Figure
3(a). The electron diffraction data indicated the presence of both hexagonal
ZnO and cubic Co_3O_4 phases, with the former predominating. However, the
individual phases could not be identified from the micrographs, which consisted
of a network of fused crystallites. Indeed, the ZnO and Co_3O_4 phases
had similar morphologies in all the samples studied. All the particle size
distributions were also similar, $\approx$ 20 nm. TEM studies of the sulphided Co/Zn
20/80 calc (Figure 3b) showed that some very distinctive changes in
morphology occurred on sulphiding. Comparison with the presulphided
Co/Zn 20/80 calc (Figure 3a) shows that the particle morphology was less
sharp-edged after sulphiding and an additional membraneous-like material
was present. Electron diffraction studies identified both ZnO and Co_3O_4 phases
as for the presulphided sample. The membraneous sheets were microcrystalline
with only weak diffuse rings in evidence in the diffraction patterns. X-ray
microanalysis of solitary areas of the membraneous material showed that it
contained cobalt, zinc and sulphur, but the ratios of these have not been
quantified at this stage. Membraneous sheets were also detected in sulphided
Co/Zn 30/70 calc and 100/0 calc, but were not found in sulphided Co/Zn
0/100 calc. ß-ZnS was identified in sulphided Co/Zn 0/100 calc, together with
ZnO, but not in any of the other samples studied, their diffraction patterns
being identical to those of the oxides.

XRD studies of the oxide phases resulting from the decomposition of the
Co/Zn/Al hydrotalcites at 723K showed that Co/Zn/Al 15/60/25 calc adopts
a ZnO type lattice whereas the diffraction patterns for Co/Zn/Al 37.5/37.5/25
calc and 60/15/25 calc corresponded to Co_3O_4. XRD studies of the sulphided
three-component systems have yet to be carried out. However, detection
of the metal sulphides may be difficult since the samples fired on removal from
the testing rig.

The surface areas of the oxides are listed in Table 4. The surface area attained
was governed by the structure of the oxide. For the Co/Zn series the surface
area generally increased with increase in cobalt concentration, decomposition
of the basic cobalt carbonate to give Co_3O_4 having the highest surface
area in the Co/Zn series. The surface areas of the three component systems were

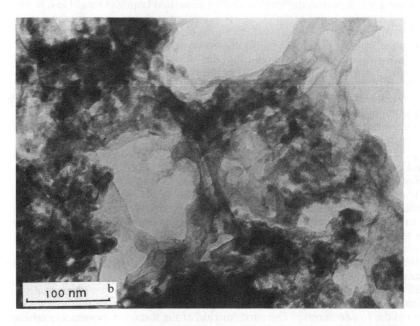

Figure 3:- TEM for (a) Co/Zn 20/80 calc., (b) Co/Zn calc and sulphided.

Table 4: Testing of Oxide Absorbents with H_2S.

Nominal Co/Zn/Al	BET Area (m^2g^{-1})	Weight (g)	Breakthrough Time (min)	% Reaction*
0/100/0	38.8	3.4998	119	13.73
10/90/0	45.1	3.9100	172	20.46
20/80/0	64.9	3.6300	209	23.86
30/70/0	68.2	2.9038	206	28.58
40/60/0	66.4	3.1259	230	33.18
50/50/0	59.7	3.8007	180.5	20.07
90/10/0	82.9	3.4489	530	63.11
100/0/0	87.1	3.4161	753	91.82
15/60/25	128.0	1.6887	83	17.8 (22.7)
37.5/37.5/25	107.9	1.5489	59	10.2 (12.6)
60/15/25	111.1	3.2212	90	13.8 (17.4)

* Calculated based on the total number of moles of H_2S absorbed at breakthrough as a percentage of the total number of moles of metal. Values in brackets are calculated based merely on moles of Co and Zn and ignoring Al.

higher than those for the Co/Zn series, the main function of the alumina being to support the Co/Zn oxides. The Co/Zn/Al 15/60/25 calc which had a ZnO type structure was found to have the highest surface area this time.

3.Absorbent testing.

The total uptake of H_2S at breakthrough for 5 cm^3 of each absorbent is listed in Table 4. The extents of reaction were determined from the ratio of the number of moles of H_2S absorbed to the total number of moles of metal in each absorbent. The extent of reaction against surface area and extent of reaction against % cobalt are plotted in Figures 4 and 5 respectively. The H_2S uptake was found to increase exponentially with surface area for the Co/Zn series, but showed little dependence on surface area in the Co/Zn/Al series. The plot of H_2S uptake against % Co showed that the absorption capacity generally increased with increase in cobalt concentration in the Co/Zn series, although the uptake was lower than expected for Co/Zn 50/50 calc. Conversely, the H_2S uptake decreased with increase in cobalt concentration in the Co/Zn/Al series.

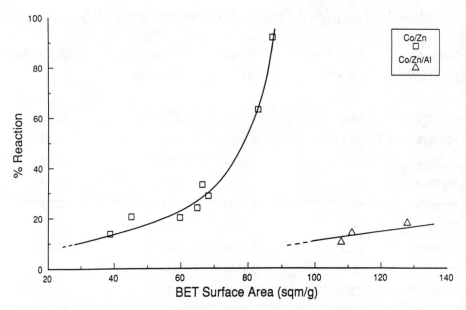

Figure 4:- BET Surface Area vs. % Reaction.

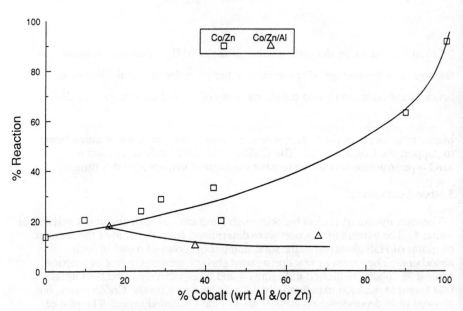

Figure 5:- % Cobalt vs. % Reaction.

Discussion

The main objective of this work was to generate high surface area absorbents for the efficient removal of H_2S at low temperatures. The initial reaction step in the absorption of H_2S by an ionic solid may be visualised as being the dissociation of H_2S into H^+ and HS^- followed by the diffusion of the HS^- into the oxide lattice and migration of oxide and water to the surface.[13,14] H_2S uptake in the Co/Zn and Co/Zn/Al samples may occur by:
 i) dissociation and subsequent absorption of H_2S on the ZnO, Co_3O_4 or both components of the mixed oxides.
 ii) preferential dissociation of the H_2S on one of the oxides followed by spillover of HS^- onto the other oxide.
Al_2O_3 would not be expected to react with H_2S, as previously stated.

The exponential increase in H_2S absorption capacity with increasing cobalt concentration in the Co/Zn series suggests that the absorption capacity shows some dependence on the structure adopted by the mixed oxides as well as their surface areas. The increase in H_2S absorption capacity with increase in cobalt concentration cannot be explained entirely by the fact that the sulphiding of cobalt oxide is more thermodynamically favoured than zinc oxide[5] since this would give a linear increase in H_2S absorption with loading. The highest uptake was for Co/Zn 100/0 calc which was formed from the decomposition of the orthorhombic $Co(CO_3)_{0.5}(OH)_{1.0}0.11 H_2O$ precursor[9] to the normal spinel Co_3O_4. The absorption capacity for this oxide was considerably greater than that for Co/Zn 90/10 calc which was formed from the decomposition of the cubic zincian spherocobaltite $Co_{0.9}Zn_{0.1}CO_3$ to zincian Co_3O_4, although the surface area only increased from 83 to 87 m^2g^{-1} on increasing the Co/Zn ratio from 90/10 to 100/0. This may be due to differences in the normal spinel in Co/Zn 100/0 calc and Co/Zn 90/10 calc arising either from the retention of some of the structural and morphological features of the precursors in the oxides[9] or the presence of zinc in solid solution influencing the oxide spinel structure in Co/Zn 90/10 calc. The influence of cobalt on the H_2S absorption capacity suggests that Co_3O_4 particles coat the ZnO particles in the mixed oxide systems. The XPS studies showed that the surface segregation of cobalt occurred in the precursors and that subsequent calcination had little effect on this.[12] This suggests that during the initial precipitation of the precursors, the hydrozincite precipitating out first and the spherocobaltite then coated the hydrozincite particles. The H_2S absorption capacities and surface areas increased only slightly with increase in cobalt concentration for Co/Zn 20/80 calc, 30/70 calc and 40/60 calc. This is not surprising since all three oxides were formed from the decomposition of hydrozincite and spherocobaltite biphasic precursors to biphasic ZnO and Co_3O_4 respectively. The XPS results indicate that some cobalt and zinc may be present in the ZnO and Co_3O_4 phases respectively. The H_2S absorption capacities for Co/Zn 10/90 calc and Co/Zn 50/50 calc were both lower than for Co/Zn 20/80 calc, 30/70 calc and 40/60 calc. In the former case, this is in accordance with the lower cobalt concentration and formation of a single phase cobaltean hydrozincite precursor which decomposed to give cobaltean ZnO. The lower uptake for Co/Zn 50/50 calc is somewhat surprising but may reflect the separation of the single phase zincian spherocobaltite precursor into ZnO and Co_3O_4 on decomposition. Co/Zn 0/100, which formed the hydrozincite precursor structure and ZnO on decomposition, was the poorest absorbent in the Co/Zn series. This was also the only sample in which the cubic β-ZnS was detected by electron diffraction after reaction with H_2S. The microcrystalline, membraneous sheets containing Co, Zn and S that have been observed in the sulphided mixed Co/Zn oxides and Co_3O_4 may be responsible for the higher H_2S absorption capacity of these samples.

The objective in the investigation of the three component Co/Zn/Al series was to prepare hydrotalcites in which all three metals were in the same crystalline structure which would form well interdispersed mixed oxides on calcination This objective was met in that hydrotalcites were formed at the three loadings studied and formed high surface area mixed oxides on calcination. The H_2S absorption capacity was poor, however, and this was probably due to the inability of Al_2O_3 to absorb H_2S. Greater solubility of cobalt in ZnO and zinc in Co_3O_4 was achieved compared with the Co/Zn series. This was probably due to the fact that the oxide was derived from the decomposition of a single hydrotalcite precursor and that a higher calcination temperature was required to effect this decomposition. It is interesting to note that in the presence of equal ratios of cobalt and zinc the zincian Co_3O_4 spinel phase was formed, reflecting the greater solubility of zinc in cobalt as also observed for the two-component oxides. The surprising result that the H_2S uptake was greatest for the cobaltean ZnO and decreased with increasing cobalt concentration again indicates that the resultant oxide is governed by the structure of the precursor. The increased intersolubility of cobalt and zinc in the three-component systems may have resulted in the formation of a higher surface area cobaltean ZnO phase with its subsequently higher H_2S absorption capacity. In addition, some of the Zn^{2+} and/or Co^{2+} ions may have been substituted into the tetrahedral sites in the alumina structure to form small amounts of x-ray amorphous metal aluminate spinels. The reasons for the lower absorption capacity of the two zincian Co_3O_4 phases are less clear. It is unlikely to be due to zinc in solid solution in Co_3O_4 as such effects were not observed for the two-component systems. The formation of small amounts of X-ray amorphous $CoAl_2O_4$ or $ZnAl_2O_4$ spinels would seem more likely. The spinels Co_3O_4, $CoAl_2O_4$, $ZnAl_2O_4$ and $ZnCo_2O_4$ are all normal spinels, so changes in spinel composition cannot be ascribed to structural changes due to partial inversion. UV/vis diffuse reflectance spectroscopic studies of Co/Zn/Al calc 37.5/37.5/25 calc and 60/15/25 also identified Co^{2+} as being in tetrahedral sites and Co^{3+} in octahedral sites, confirming the assignment of these oxides to the normal spinel configuration.[15] However, the magnetic properties of the spinels would be affected by ion substitution and may contribute to the observed effects on H_2S absorption capacity. Further work is now underway to investigate the intersolubility of the mixed oxides.

Conclusions

1. The H_2S absorption capacity increases with increasing cobalt content in the Co/Zn series. This is associated with a change in structure from hexagonal ZnO to cubic Co_3O_4, an increase in surface area and a surface excess of cobalt ions.

2. The three component oxides originating from the hydrotalcite precursors had higher surface areas, but were poorer absorbents owing to the inability of aluminium to react with H_2S. The Co/Zn/Al hydrotalcite that adopted the ZnO structure on calcination was the best absorbent in this case. $ZnCo_2O_4$ and/or $CoAl_2O_4$ spinels may also be present.

3. ZnS was detected only in sulphided ZnO. Microcrystalline sheets containing Co, Zn and S were identified in the sulphided cobalt and zinc mixed oxides.

Acknowledgments

We are grateful to SERC and ICI Katalco for supporting this work.

References

1. C.H. Bartholomew, P.K. Agrawal and J.R. Katzer; Adv. Catal, 1982, 31, 135.

2. P. O Neill, "Environmental Chemistry," Chapman and Hall, Second Edition, London, 1993.

3. P.J.H. Carnell, in "Catalyst Handbook," ed. M.V. Twigg, Wolfe Publishing Ltd, London, 1989, p191.

4. P.J.H. Carnell and P.E. Starkey, Chem. Eng., 1984, 408, 30.

5. T. Baird, P.J. Denny, R. Hoyle, F. McMonagle, D. Stirling and J. Tweedy; J. Chem. Soc. Faraday Trans., 1992, 88, (22), 3375.

6. F. Cavani, F. Trifiro and A. Vaccari,Catal. Today; 1991, 11,(2), 173.

7. K. Petrov, L. Markov, R. Ioncheva and P. Rachev; J.Mater. Sci., 1988, 23, 181.

8. P. Courty, D. Durand, E. Freund and A. Sugier; J. Mol. Catal., 1982, 17, 241.

9. P. Porta, R. Dragone, G. Fierro, M. Inversi, M.L. Jacano and G. Moretti; J. Mater. Chem.; 1991, 1, (4), 531.

10. W.T. Reiche; Solid State Ionics; 1986, 22, 135.

11. P. Porta, R. Dragone, G. Fierro, M. Inversi, M. Lo Jacano and G. Moretti; J. Chem. Soc. Faraday Trans.; 1992, 88, (3), 311.

12. R. Hoyle, T Baird, K.C. Campbell, P.J. Holliman, M. Morris, D. Stirling and B.P. Williams; results currently being submitted for publication.

13. C.H. Lawrie; PhD Thesis, Edinburgh, 1991.

14. P.J. Holliman, R. Millar; unpublished results.

15. P.J. Holliman; unpublished results.

A Novel Technique To Measure *in situ* Wettability Alterations Using Radioactive Tracers

R. N. Smith and T. A. Lawless

AEA TECHNOLOGY, PETROLEUM SERVICES, WINFRITH, DORCHESTER, DORSET DT2 8DH, UK

INTRODUCTION

The ability to obtain representative laboratory core analysis data is influenced both by the measurement technique employed and the state of the samples used (1). The act of cutting core samples, bringing them to the surface and storage, can change the wetting state of the sample from that existing in the reservoir (2). Although the effect of wettability on some core analysis measurements is well known, uncertainties associated with reproducing the field wettability have meant that tests have often been carried out with cleaned cores, which may have an unrepresentative water wet state (1). Although wettability can be altered in the laboratory it is difficult to know whether the field wettability has been restored (3).

Many methods have been evaluated to measure wetability. Anderson (3) has described a number of quantitative and qualitative methods currently in use. Although no single accepted method exists, qualitative techniques are the most widely used. The contact angle technique measures the wettability of a specific surface, whilst the Amott (imbibition and forced displacement) and US Bureau of Mines methods measure the average wettability of a core.

Due to the difficulties and inadequacies associated with the measurement of wettability in cores, the development of an in-situ method would be of great benefit. Ferreira et al (4) have reported the results of a simulated single-well back flow tracer test to estimate in-situ reservoir wettability. Reservoir wettability has been inferred from production and tracer data, which had varied as a result of variations in the wetting condition of the formation and thus affected the transport properties of the reservoir.

Described in this paper are the experimental routines and results from a series of tests conducted to assess the potential of a novel, non-destructive technique designed to quantify wettability alteration. The technique involves the use of radioactive tracers to describe the degree of ion exchange in a variety of lithologies.

It is believed that by understanding ion-exchange processes it may be possible to establish the core wettability from chemical data and develop a technique, which may be applied to defining, in-situ, the wettability of a reservoir.

BACKGROUND

Wettability

The assessment of recoverable oil reserves, via water injection projects, is dependent, in part, upon the accurate determination of relative permeability, residual oil saturation and resistivity data. Based upon production data from existing fields, it has been shown that very often the early special core analysis data used in field development studies were inaccurate and in some cases had significant implications for development plans and costs.

It is now generally recognised that wettability is a major factor in the determination of relative permeability and residual oil saturation. Much of the early data is probably erroneous, as the correct wettability was not established in the laboratory and inappropriate criteria were applied in the design of displacement tests.

Wettability, as applied to an oil reservoir, describes the tendency of a fluid to adhere or absorb to a solid surface in the presence of another immiscible fluid. It can be described as a measure of the affinity of the rock surface for the oil or water phase. A major role of wettability in a reservoir is that of determining the location and distribution of reservoir fluids that influence relative permeabilities and thus recovery efficiency. Wettability is thus a major factor in determining the degree of oil recovery from a reservoir (5-8). The amount of oil recovery, as a function of water injection, is dependent upon the wetting state of the reservoir; thus, the evaluation of reservoir wettability is critical in the determination of specific production processes (9-11).

The conditions that establish a given reservoir wettability are not well known. The fluid movement through a reservoir, temperature and pressure changes, fluid production and the injection of fluids and chemicals used to enhance production are all factors that must be considered as affecting wettability. Research has indicated that surface active components of crude oil can be important in defining reservoir wettability (12-14).

The properties of reservoir rock are also factors in determining wettability. Significant variations in wettability may be related to variations in pore surface texture and mineralogical composition. The presence of water or previously adsorbed organic films, possibly from contact with crude oil or other organic materials, is an additional factor that can influence wettability (5).

Only a fraction of the constituents of crude oil are believed to be capable of reacting with the reservoir rock surface. Several researchers have indicated that the wettability of a reservoir is strongly related to the amount of adsorption by the heavy ends found in oil (15-18). The heavy ends contain the most polar class of compounds found in the crude oil and are principally asphaltene and resin fractions (17).

One approach to gain an insight into wettability, has been adsorption studies of crude oil and oil components on reservoir rock and minerals. Adsorption of heavy ends onto clay minerals has been reported in the literature (16). Adsorption was found to depend on the cationic form of the clay and on the solvent used for heavy ends dissolution. Subsequent work found that the adsorption of asphaltenes onto clays and minerals was reduced by the presence of water (17).

Improving oil production from a reservoir depends on a good, fundamental understanding of the interaction that occurs between the reservoir fluids and the reservoir matrix. Wettability and adsorption studies are a means to increase this understanding.

Cation Exchange Capacity

Due to their large surface area with unsatisfied native charges, clay minerals found in rocks are very reactive. Ions in solution are easily absorbed/desorbed from clays, making them very important in the process of ion exchange. The effect of ion exchange in clays on bulk water composition in a closed environment has been demonstrated to be significant. Conversely, however, it is clear that water chemistry can be used to control the exchange process (19). Clays occur in reservoirs as pore filling/grain rims and can seriously affect reservoir properties (eg porosity and permeability).

The affinity of an ion towards a given ion exchanger, ie the ion exchangeability, depends primarily on the electrical charge of the ion, the ionic radius, its relative abundance and the degree of hydration. The larger the charge on the ion, the greater is its exchange capacity (5). In the case of equivalent ions, the magnitude of their radii is decisive in their exchange capacity. The greater the volume of the ion, the weaker is its electric field in solution and thus the smaller its degree of hydration. The hydrodynamic radii of ions are seen to decrease with increasing atomic weight and hence their exchange energy (the energy with which the ion is transported from the solution into the ion exchanger) increases. Thus the exchange capacity of cations is inversely proportional to the hydrated ion radius. Note, however, that the degree of ion hydration depends on solution concentration, temperature and the presence of competing ions.

Ions can be arranged into a series, according to their exchange energy. The order of ions in the series will depend upon the properties of the solution (pH, concentration, nature of the solvent), however a generalised cation series is shown below :

$$Na^+ < NH_4^+ < K^+ < Sr^{2+} < Cs^+ < Mg^{2+} < Ca^{2+} < Cd^{2+} < Co^{2+} < Al^{3+} < Fe^{3+}$$

This arrangement, in order of their exchange efficiency, is termed the Lyotropic or Hofmeister series.

Ion exchange occurring on a cation exchanger RMe_2 can be represented by the formula :

$$RMe_2 + Me_1X \qquad\qquad RMe_1 + Me_2X$$

where R is the ion exchange media with the functional group, Me_2 is the mobile cation liable to exchange, and Me_1X is the electrolyte in solution.

Ion exchange is therefore a reversible reaction which, depending upon the concentration, properties of the ions and nature of the exchanger, can proceed in either direction.

EXPERIMENTAL - MATERIALS AND METHODS

Tracers

In North Sea production operations, seawater is routinely employed in offshore oilfields for water injection (secondary recovery) to maintain pressure and afford a sweep pattern for the mobilization of oil to the production wells. The chemical composition for seawater varies in different parts of the world. Table 1 shows the major components of North Sea brine; within this region there is very little variation of components (except perhaps near the estuaries of major rivers). Relative to formation water there is, as a general rule, less NaCl and a smaller range of cations.

Based on knowledge of the relative abundance of the cations in seawater and of the Lyotropic or Hofmeister series, two radio-labelled cations, namely

Table 1

Typical Major Components Of North Sea Sea Water*

Sodium (Na)	11000
Potassium (K)	460
Calcium (Ca)	476
Magnesium (Mg)	1440
Barium (Ba)	0.1
Strontium (Sr)	7
Iron (Fe)	0.05

Chloride (Cl)	19000
Sulphate (SO_4)	2725
Carbonate (CO_3)	NIL
Bicarbonate (HCO_3)	145
Hydroxyl (OH)	NIL
Total Dissolved Solids	35000
pH	7.8

NOTE:

All data as mgl^{-1} except pH·

* From Johnson K.S., "Water Scaling Problems in the Oil Industry". Royal Soc. Chem.

[45]Ca and [22]Na, were selected for use in this study. In addition, a neutral tracer (tritium) was selected, in order that the exchange of sodium and calcium ions could be measured relative to the non-interacting water tracer.

Geological Materials

A suite of rock types were selected to cover a wide range of mineralogies and potential wettability states. Four lithologies have been selected; three sandstones (Clashach, Rosebrae and Lochabriggs) and one limestone (Portland).

Clashach and Portland are very pure examples of sandstone and limestone respectively, whilst the two other lithologies, Rosebrae and Lochabriggs, are sandstones known to have clay mineral contents of 5-6 wt% (see Appendix 1 for mineralogical details).

In general terms sandstones can be regarded as water-wet ie water coats the grains of the rock and may fully occupy the smaller pores. Limestones are commonly believed to be oil-wet, where the reverse of the above is prevalent and oil coats the grains and occupies the smaller pore spaces.

Coreflooding Protocol

An extensive testing matrix was devised, to examine the retardation (and hence ion exchange) of sodium and calcium (carried in seawater from the Chesil beach or NaI brine) in the four selected lithologies, in the absence and presence of oil.

Retardation effects (and ion exchange) are considered to be a direct consequence of rock-fluid interactions and may be correlated to core wettability. Any change in such retardation brought about by, for example, prolonged contact with oil may thus be established and the resulting wettability inferred.

The following test sequence was invoked for all corefloods:

(i)   A 6" long by 1.5" diameter core was housed in a core flood assembly, fitted with pumps and transducers (see Figure 1). The core holder was located in a constant temperature bath set at 30°C.

(ii)   The core was initially flooded with untraced seawater (filtered to 0.45 microns) for 10 pore volumes before establishing the absolute permeability to seawater.

(iii)   Four pore volumes of traced seawater (tritium plus either [45]Ca or [22]Na) were injected into the core at 100 cm^3h^{-1} (or as specified in the following sections).

(iv)   Immediately after the injection of traced seawater, 8 pore volumes of untraced seawater were injected at 100 cm^3h^{-1}.

(v)   From stages (iii) and (iv) the fluids eluted from the core were collected in 5 cm^3 aliquots.

(vi)   From each aliquot, 0.5 cm^3 samples were taken and added to 4.5 cm^3 of demineralised water and 5 cm^3 of a scintillation "cocktail" and placed in a glass vial. The sample vials, plus standards, were then analysed using a Beckman beta-counter.

vii)   The counts for interactive and non-interactive tracers were background corrected and normalised against standards. For each experiment the normalised counts for each tracer were plotted against the number of pore volumes eluted from the core. In this manner the retardation of the ^{45}Ca or ^{22}Na due to ion exchange was established.

RESULTS AND DISCUSSION

The Effects of Lithology

Adsorption/desorption profiles for the two tracers, ^{45}Ca and ^{22}Na, in each of the four lithologies are shown in Figures 2 to 9.

Examining the profile of the non-interactive isotope (tritium) in each core flood it is clear that the levels of tritium rise steadily as traced seawater replaces the water originally resident in the core (the dispersion profile). After injection of two pore volumes the core is fully saturated with traced seawater and the normalised counts remain at unity until untraced seawater is injected. The observed counts then fall back, effectively to zero, as untraced water replaces the traced water in the core. Note, if no ion exchange processes occur within the core, then the profiles for ^{45}Ca and ^{22}Na would follow the same path to that recorded for tritiated water. Any delay/retardation of the labelled cation, with respect to the tritium, indicates the magnitude of the cation exchange process.

With regard to sodium, there appears to be little or no delay/retardation in Clashach, Rosebrae or Portland (see Figures 2, 4 and 8 respectively). Clashach and Portland are both very pure rocks with clay minerals found at trace levels only. In addition, monovalent labelled ^{22}Na will face competition for limited ion-exchange sites from sodium and divalent ions found naturally in the carrier seawater. There is a small delay for ^{22}Na, with respect to tritium, in the Rosebrae sandstone. However, the largest delay is in Lochabriggs sandstone (see Figure 6), although again retardation is not particularly marked.

For calcium the retardation is greater in all tested lithologies, as shown by examination of Figures 3, 5, 7 and 9, with the most significant delay being recorded for the Lochabriggs sandstone (Figure 7).

In Portland, the delay of ^{45}Ca is considerable (see Figure 9). In this limestone (composed almost entirely of calcium carbonate) there are many sites

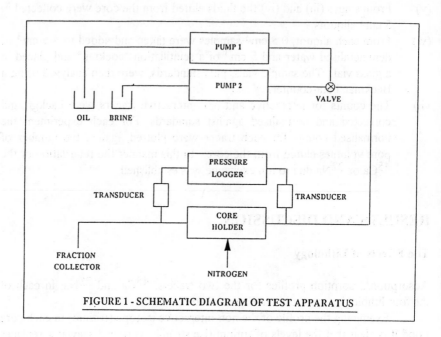

FIGURE 1 - SCHEMATIC DIAGRAM OF TEST APPARATUS

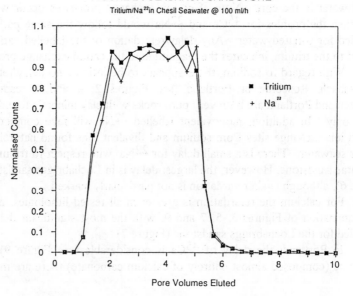

FIGURE 2 : CLASHACH SANDSTONE

Tritium/Na^{22} in Chesil Seawater @ 100 ml/h

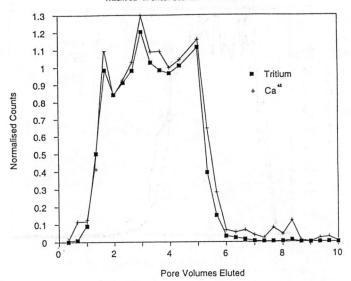

FIGURE 3 : CLASHACH SANDSTONE

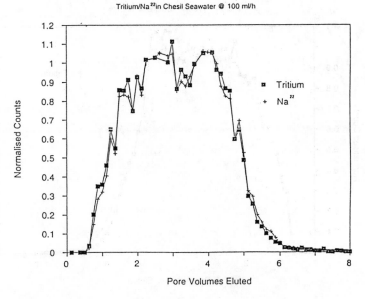

FIGURE 4 : ROSEBRAE SANDSTONE

FIGURE 5 : ROSEBRAE SANDSTONE

Tritium/Ca ^{45} in Chesil Seawater @ 100 ml/h

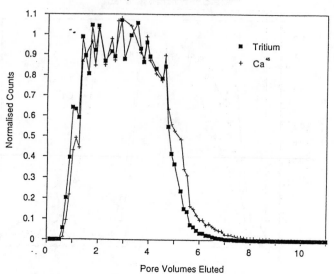

FIGURE 6 : LOCHABRIGGS SANDSTONE

Tritium/Na ^{22} in Chesil Seawater @ 100 ml/h

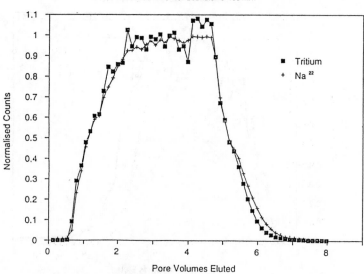

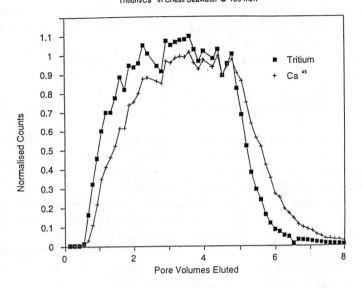

FIGURE 7 : LOCHABRIGGS SANDSTONE
Tritium/Ca^{45}in Chesil Seawater @ 100 ml/h

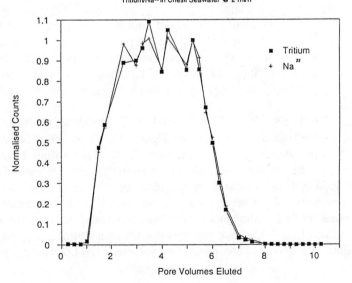

FIGURE 8 : PORTLAND LIMESTONE
Tritium/Na^{22}in Chesil Seawater @ 2 ml/h

for the calcium to absorb onto. Note that due to the low permeability of the core, it was necessary to inject fluids at only 2 cm^3h^{-1}. With such low flow rates, coupled with the abundance of available adsorption sites, it appears that ^{45}Ca was being adsorbed and stripped, then re-adsorbed numerous times over the length of the core. For this core and at the utilised flow rate of 2 cm^3h^{-1}, more than 20 pore volumes would be required to propagate the ^{45}Ca through the core.

The Role of the Carrier Media

For the experiments described above Chesil seawater (filtered to 0.45 microns) was used as the carrier fluid for the tracers. For each of the lithologies the retardation of ^{45}Ca (with a carrier loading of 50 mgl^{-1} Ca^{2+}) in 3% w/w NaI brine has also been examined (see Figures 10 to 13).

NaI brine was evaluated because of its use in gamma attenuation experiments, and is an integral component of existing in-situ wettability and saturation monitoring techniques performed within our laboratories.

For these experiments the test regime detailed above was adopted, except for item (iii). At this stage four pore volumes of traced brine (made up from 3% w/w NaI and 50 ppm $CaCl_2$ in demineralised water) were injected into the core at 100 cm^3h^{-1} for the sandstones and 3 cm^3h^{-1} for the Portland limestone. The fluid resident in the core prior to injection of NaI and the desorbing fluid in each test was Chesil seawater.

For all four lithologies the effect of NaI on retardation is considerable. In Clashach sandstone (see Figure 10), ^{45}Ca is strongly retarded, with respect to the tritiated water. However, during tracer displacement the Chesil seawater immediately flushed out a large proportion of the bound ^{45}Ca (note the peak at 5.5 pore volumes). Continued injection of seawater allows the counts to return to background levels after approximately eight pore volumes. It is apparent that a large proportion of the calcium ions are being removed from solution in preference to the Na^+ ions. However, on reverting to seawater injection the ^{45}Ca ions are readily displaced by the divalent and monovalent ions in the seawater.

A similar phenomenon is observed in the Rosebrae sandstone, with a rapid release of ^{45}Ca from the rock (see Figure 11) on injection of seawater.

For the Lochabriggs sandstone (see Figure 12), a strong adsorption of ^{45}Ca is observed. However, on injecting seawater the rapid release of calcium, seen in Rosebrae and Clashach, is not observed. What is seen is a gradual increase in counts up to seven pore volumes, with a gradual decrease back to background levels. Although this test was terminated at 12 pore volumes, extrapolation of the ^{45}Ca trace indicates that the final removal of active calcium ions would be observed at around 14 pore volumes. The Lochabriggs sandstone

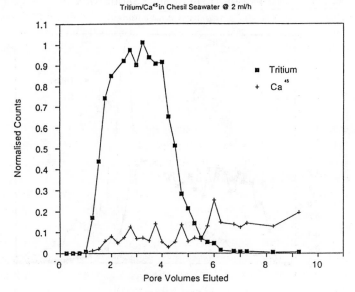

FIGURE 9 : PORTLAND LIMESTONE

Tritium/Ca^{45} in Chesil Seawater @ 2 ml/h

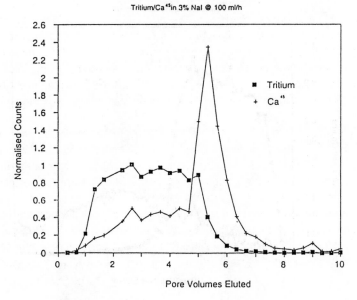

FIGURE 10 : CLASHACH SANDSTONE

Tritium/Ca^{45} in 3% NaI @ 100 ml/h

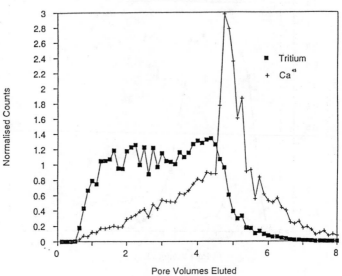

FIGURE 11 : ROSEBRAE SANDSTONE

Tritium/Ca^{45} in 3% NaI @ 100 ml/h

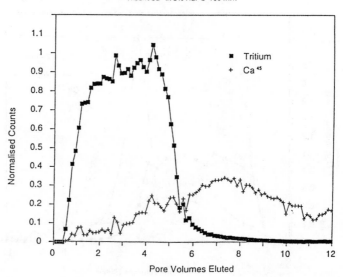

FIGURE 12 : LOCHABRIGGS SANDSTONE

Tritium/Ca^{45} in 3% NaI @ 100 ml/h

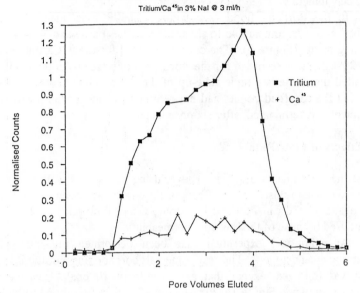

FIGURE 13 : PORTLAND LIMESTONE

Tritium/Ca^{45} in 3% NaI @ 3 ml/h

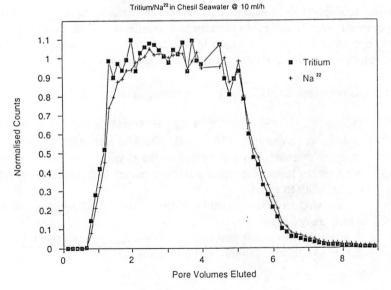

FIGURE 14 : LOCHABRIGGS SANDSTONE

Tritium/Na^{22} in Chesil Seawater @ 10 ml/h

is known to have a different clay mineral content to the Rosebrae and Clashach rocks and it is apparent that the brines used in these tests have different effects on the clay minerals.

In the Portland flood (see Figure 13), ^{45}Ca is again very strongly adsorbed at levels comparable to those where Chesil seawater was used as the desorbing media (Figure 9). The counts recorded for the ^{45}Ca only ever reach about 20% of those recorded for the stock (non-injected) solution and it must be concluded that there is a rapid take-up of the ^{45}Ca, with very slow desorption (note that the recorded counts had not returned to background levels when the experiment was terminated, after six pore volumes).

The Effects of Flow Rate

For all the experiments described above, using sandstones, a flow rate of 100 cm^{3}h^{-1} has been utilised. The effects of flow rate have been alluded to earlier with regard to ^{45}Ca in Portland limestone, where flow rates of 2 cm^{3}h^{-1} had been utilised because of the low core permeability.

An additional experiment has been performed utilising ^{22}Na and Lochabriggs sandstone. The experimental coreflooding protocol described earlier was followed, except that a flow rate of 10 cm^{3}h^{-1} was employed. Results are shown in Figure 14. Retardation of sodium was found to be less than one-third of a pore volume behind the tritium and is comparable with data recorded for this lithology and ^{22}Na at flow rates of 100 cm^{3}h^{-1} (Figure 6).

The Effects of Oil

Having established the retardation of tracers in the absence of oil, a series of experiments were devised to determine the effects of oil presence on ion exchange processes. Any deviation in retardation would be due to exclusion of the water phase and/or alteration of the rock wettability.

Three oils have been used for these experiments, namely :

i)   Decane : a pure mineral oil, chosen to provide a base case. As the oil contains no impurities, which could alter the formation wettability, any changes in retardation will be due to the physical presence of oil in the pore spaces restricting water phase occupancy and reducing the relative permeability to water.

ii)   A low viscosity (stock tank) oil; the wettability altering properties of which are unknown.

iii)   A high viscosity (stock tank) oil. From previous studies this oil has been shown to alter the wettability of outcrop materials.

The following experimental sequence was invoked for this section of the programme.

i)   A 6" long by 1.5" diameter core was housed in a core flood assembly, fitted with pumps and transducers (see Figure 1).

ii)   The core was then flooded with untraced seawater for 10 pore volumes, before establishing its permeability to seawater.

iii)   Decane was injected into the core until residual water saturation was achieved. The permeability to decane at residual water was then determined.

iv)   Chesil seawater was injected into the core until residual decane saturation was achieved. The permeability to seawater at residual decane was then established.

v)   Four pore volumes of traced seawater was then injected into the core at 100 cm^3h^{-1}.

vi)   Immediately after the injection of traced seawater (containing tritium and ^{45}Ca), 8 pore volumes of untraced seawater were injected at 100 cm^3h^{-1}.

vii)   From stages (v) and (vi) the fluids eluted from the core were collected in 5 cm^3 aliquots.

viii)   Crude oil was then injected into the core until residual water saturation was achieved. The permeability to oil at residual water was then established.

ix)   The final pore volume of oil was allowed to remain in-situ in the core for 72 hours.

x)   Repeat (iv) to (vii) above.

xi)   From each aliquot a 0.5 cm^3 sample was taken and added to 4.5 cm^3 of demineralised water and 5 cm^3 of a scintillation "cocktail" and placed in a glass vial. The sample vials, plus standards, were then analysed using a Beckman beta-counter.

xii)   The counts for interactive and non-interactive tracers were background corrected and normalised against standards. The standards being two samples taken from a stock solution that had not been injected into the core. For each experiment the normalised counts for each tracer have been plotted against the number of pore volumes eluted from the core.

For these experiments the Lochabriggs (see Figures 15 and 16) and Rosebrae sandstone (see Figures 17-19) were utilised with the ^{45}Ca and tritium tracers.

     In quantifying retardation effects, a delay (in terms of volume throughput) can be defined for the labelled cation, with respect to tritium at the point where $C/C_0 = 0.5$. This is the ratio of eluate concentration C to input concentration C_0. (For a non-interactive tracer, such as tritium, $C/C_0 \approx 0.5$ should be at one pore volume).

With regard to the Lochabriggs sandstone, in the absence of oil a delay for the ^{45}Ca, with respect to tritium, of 0.44 pore volumes on injection was observed (Figure 7). Delays of 0.25 and 0.13 pore volumes have been recorded in the presence of decane and the low viscosity crude. These results indicate that the presence of oil in the pore spaces may reduce the ion-exchange potential of the rock.

For the Rosebrae sandstone, a delay of 0.15 pore volumes was seen for ^{45}Ca in the absence of oil (see Figure 5). The retardation of ^{45}Ca, with respect to tritium, in the presence of decane and both crude oils was increased to 0.3 pore volumes. It is apparent that these oils are having some affect, albeit small.

From the results of this initial study it is possible to visualise the potential of this technique to quantify wettability alteration. Note, at this stage, no independent tests have been conducted to establish whether or not the selected crude oils would alter the wettability of Rosebrae sandstone in 72 hours at 30°C. However, one would expect that if the rock's internal surface was oil wetted then any subsequent interaction with a flowing aqueous phase would be impossible and as a result any retardation effects will be absent.

CONCLUSIONS

A suite of core flooding experiments have been performed to establish the retardation (ion exchange) of sodium and calcium tracers, in four selected lithologies, in the absence and presence of oil.

The following conclusions may be drawn from the experimental work:

i)   For seawater solutions there is little or no delay of ^{22}Na in samples of pure sandstone and limestone. There is only a small retardation in the Lochabriggs sandstone. This sandstone is known to contain the clay minerals sericite, illite and kaolinite.

ii)   There is a much more pronounced retardation of ^{45}Ca, with respect to tritium, in all the tested lithologies. Significant delays have been observed in the clay bearing Rosebrae and Lochabriggs sandstones.

In the Portland limestone, the many adsorption sites available for calcium ions, coupled with the low injection rates, has resulted in a considerable retardation of the injected ^{45}Ca ions. It is suggested that ^{45}Ca is being repeatedly adsorbed and desorbed many times along the length of the

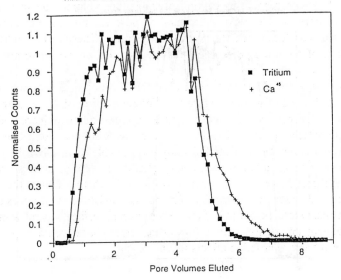

FIGURE 15 : LOCHABRIGGS SANDSTONE

Tritium/Ca^{45}in Chesil Seawater @ 100 ml/h in the Presence of Decane

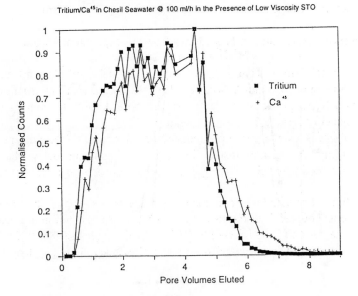

FIGURE 16 : LOCHABRIGGS SANDSTONE

Tritium/Ca^{45}in Chesil Seawater @ 100 ml/h in the Presence of Low Viscosity STO

FIGURE 17 : ROSEBRAE SANDSTONE

Tritium/Ca^{45} in Chesil Seawater @ 100 ml/h in the Presence of Decane

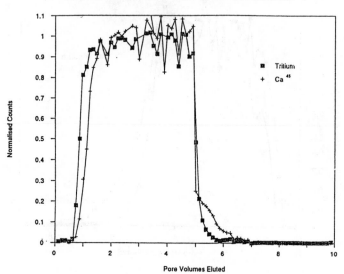

FIGURE 18 : ROSEBRAE SANDSTONE

Tritium/Ca^{45} in Chesil Seawater @ 100 ml/h in the Presence of Low Viscosity STO

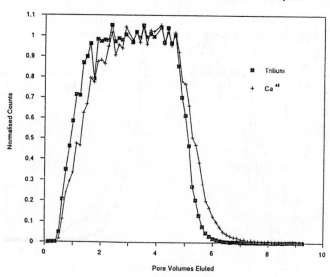

FIGURE 19 : ROSEBRAE SANDSTONE

Tritium/Cd^s in Chesil Seawater @ 75 ml/h in the Presence of High Viscosity Crude

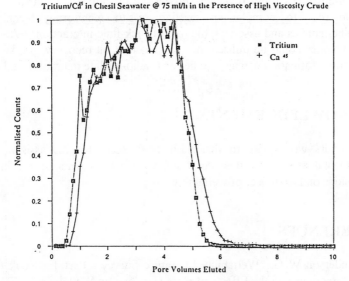

core and at the flow rates used in these experiments, > 20 pore volumes of untraced seawater would be required to desorb all the ^{45}Ca.

iii)   In all four lithologies it has been observed that the use of 3% NaI brine, as the carrier media for the tritium and ^{45}Ca tracers, has a considerable effect on the ion exchange processes. On reverting to injection of seawater, with the Rosebrae and Clashach sandstones, the clays are returned to their native/non-active state and the ^{45}Ca ions are rapidly exchanged.

iv)   In one test with Lochabriggs sandstone and ^{22}Na, a reduction in flow rate from 100 to 10 cm^{3}h^{-1} was observed to have no affect on the ion exchange process.

v)   In experiments performed at residual decane and low viscosity oil saturation, with Lochabriggs sandstone, the presence of oil was found to reduce the ion exchange potential of the rock.

vi)   In experiments performed using the Rosebrae sandstone at residual decane and crude oil saturation, a slight increase in the retardation of ^{45}Ca was observed. The viscous crude has known wettability altering properties, however it is unknown whether or not the oil could alter the wettability of Rosebrae sandstone in 72 hours at 30°C.

Despite the result reported in (vi) above, it is envisaged that the development of this non-destructive and reversible process in quantifying wettability will be a valuable tool to complement existing in-situ monitoring techniques for wettability studies and assessing the role of rock-fluid interactions. Areas where the techniques could be utilised include enhanced oil recovery, produced water re-injection, formation damage studies and radio-nuclide migration studies.

ACKNOWLEDGEMENTS

The authors would like to thank Mr S P Beare for his assistance in the experimental aspects of this work and to Mr D A Puckett for his helpful discussions on the subject of wettability.

REFERENCES

1. Anderson W G, "Wettability Literature Survey - Part 1 : Rock /Oil/Brine Interactions and the Effects of Core Handling on Wettability". JPT October 1986.

2. Wendel D J, Anderson W G & Meyers J D, "Restored-State Core Analysis for the Hutton Reservoir". SPE Formation Evaluation, Dec 1987.

3. Anderson W G, "Wettability Literature Survey - Part 2 : Wettability Measurement". JPT November 1986.

4. Ferreira L E A, Descant F J, Mojdeh Delshad, Pope G A & Kamy Sepehrnoori, "A Single-Well Tracer Test to Estimate Wettability". SPE/DOE 24136.

5. Crocker M E and Marchin L M, "Wettability and Adsorption Characteristics of Crude Oil Asphaltenes and Polar Fractions". JPT April 1988.

6. Lorenz P B, Donaldson E C & Thomas R D, "Use of Centrifugal Measurements of Wettability to Predict Oil Recovery", USBM Report of Investigations (1974) 7873.

7. Donaldson E C, Thomas R D & Lorenz P B, "Wettability Determination and Its Effect on Recovery Efficiency", SPEJ (March 1969) 13-20.

8. Kinney P T & Nielsen R F, "Wettability in Oil Recovery". World Oil (April 1951) 145.

9. Emery L W, Mungan N and Nicholson R W, "Caustic Slug Injection in the Singleton Field" JPT (Dec 1970) 1569-76.

10. Michaels A S & Porter M G, "Water Oil Displacements from Porous Media Utilizing Transient Adhesion-Tension Alterations", AIChE J (July 1965) 617-24.

11. Leach R O et al, "A Laboratory and Field Study of Wettability Adjustment in waterflooding" JPT (Feb 1962) 206-12, Trans AIME 225.

12. Bartell F E & Niederhauser D O, "Fundamental Research on Occurrence and Recovery of Petroleum 1946-47, API Washington DC (1949) 57-80.

13. Dodd C G, Moore J W & Denekas M O, " Metalliferous Substances Adsorbed at the Crude Petroleum-Water Interfaces", Ind. Eng. Chem (1952) 44, 2585-90.

14. Seifert W K & Howells W G, " Interfacially Active Acids in a California Crude", Anal. Chem (Oct 1969) 41 1638-47.

15. Donaldson E C & Crocker M E, "Characterisation of the Crude Oil Polar Compound Extract", DOE Report of Investigations, DOE/BETC/RI-80/5, US DOE (Oct 1980).

16. Clementz D M, "Interaction of Petroleum Heavy Ends with Montmorillonite" Clays and Clay Minerals (June 1976) 24 312-19.

17. Collins S H & Melrose J C, "Adsorption of Asphaltenes and Water on Reservoir Rock Minerals", Paper SPE 11800 presented at the 1983 SPE Intl. Symposium on Oilfield and Geothermal Chemistry, Denver, June 1-3.

18. Cuiec L, "Rock/Crude-Oil Interactions and Wettability: An Attempt to Understand Their Interrelation", Paper SPE 13211 presented at the 1984 SPE Annual Technical Conference and Exhibition, Houston, Sept 16-19.

19. Scheuerman R F and Bergersen, B M, SPE 18461, 1989.

Appendix 1 : Details of Selected Lithologies

A1.1   Clashach Sandstone

Clashach is a well-sorted, equigranular, supermature quartz-arenite. Clasts are almost entirely well rounded quartz with approximately 5% K-feldspar cemented by silica. Small quantities of illite and muscovite are also present. The rock has a porosity of around 18% and a grain density of 2.64 g/cm^3.

A1.2   Rosebrae Sandstone

Rosebrae is a very fine grained rock with moderate sorting and can be described as a sub-arkosic sandstone. The rock is made up of around 86% quartz and 8% K-feldspar. Rosebrae contains considerably more clay than Clashach sandstone (around 6%). Species observed are illite and smectite. The rock has a porosity of around 23% and permeability of approximately 200 md.

A1.3   Lochabriggs Sandstone

Lochabriggs is a sub-arkosic sandstone, very fine to medium grained, well sorted with sub-angular to sub-rounded grains. The rock is made up of around 76% quartz and 11% feldspars, however most of the feldspars are partially or completely altered to sericite, illite or kaolinite. Moderately developed illite rims coat the framework grains of the rock (around 4 wt% of the bulk rock). Porosities in the region of 22% have been recorded, with permeabilities between 200 and 300 md.

A1.4   Portland Limestone

Portland limestone is a high purity limestone with only minor contamination by quartz, and possibly dolomite. The rock can be described as an oobiospararenite. A porosity value of around 16% has been obtained for the rock, with a liquid permeability in the region of 0.5 md.

Oil and Gas Production: Future Trends of Relevance to the Oilfield Chemical Supply Industry

R. W. Johnson

THE PETROLEUM SCIENCE AND TECHNOLOGY INSTITUTE, EDINBURGH, UK

1.0 Introduction

The purpose of this paper is to give an overview of some current trends in the oil and gas industry which are likely to impact the oilfield chemical industry in the medium (3-8 year) term. The impact of these trends is already starting to be felt in the industry to a greater or lesser extent. The premise here is that the chemicals industry stands at the early part of an S-curve of change; that the rate of change of business environment within the oil industry - particularly in Europe- will gather pace in the next 2-3 years; that recent 'restructurings' in the industry represent but the beginning of this change; that the forthcoming shape of the industry will provide significant opportunities for some companies in the chemical supply industry to become leaders in the application of new product and process technology; and that companies who recognise the opportunities will survive, whilst those that do not will have a limited future.

This is primarily a broad brush technology overview paper. However, because of the dominance of oil price economics, this issue cannot be ignored as it conditions the overall business outlook.

2.0 Industry background conditions

Overall there are three primary forces at play. These forces are coupled and interwoven in a complex way, and cannot be regarded as independent. Changes in any one affect and impact the other two. The forces are as follows:-

2.1 Economic

The oil industry is currently operating in a market where supply overhangs demand. The pressure on the price of oil is therefore primarily downward. The likelihood of companies planning, long term, on prices outside the $12-$15/barrel range is regarded by many accepted experts as small. A depressed price (depressed that is relative to recent years, rather than the 'historic' price) has two effects:-

(i) restriction of margins which in turn reduces high risk/high cost exploration activity, and the addition of new reserves. As the major proportion of oilfield chemical use (Ref 1) is drilling-related this is the main factor which determines the overall market for chemicals.

and

(ii) assurance of the dominance of oil and gas as the world's primary energy source. Until another similarly cheap, and flexible, source of primary energy is available the long term future of the industry as a whole is not an issue.

This is not the same as the long term future of any single *company* within the industry. Companies --both operators and suppliers-- who find themselves incorrectly positioned in terms of the current economic climate may not survive in their existing form.

Lower energy prices will undoubtedly help newly industrialising countries in their efforts to develop. As they do so, their energy demand will rise. This is particularly likely in the Pacific Rim. The rate of growth in demand may be such that it stretches the ability of the oil industry to satisfy it out of shut-in capacity plus output from new discoveries which result from a reduced level of exploration. If this is the case then prices will rise. This aspect is generally seen as the 'unknown' in the price equation. Rightly, few companies are betting their future on such a price rise, though a number may be positioning themselves to take advantage if ths should happen.

One main feature of the recent continuous slide in oil price has been the impetus it has given to reductions in the cost base in high cost offshore production in the UK and Norway. This reduction has been achieved in the short term through shrinking of internal staffing on the part of operating companies, on simplification of hardware design for field application and the detailed investigation of life cycle production costs. A typical example of the latter kind of activity which oil companies are now pursuing is detailed examination of the cost advantage of high spec materials, or benefits of working at different operating conditions, versus use of corrosion or other types of inhibitors.

Reduction in internal manpower opens, indeed forces, dependence on service companies. Strongly competent, innovative, high added value companies, are more likely to thrive in these business conditions and to do so at the expense of their competitors.

2.2 Environmental

Operational environmental considerations loom large as a primary business driver. This has been the case for the industry for at least the last 25 years; the difference is that what was previously a serious, but understated, activity has now become a high profile foreground consideration.

Oil companies were proactively and openly addressing environmental policy issues in the late sixties and early seventies (Refs 2,3), and have worked to minimise the impact of the industry in relation to the environment by anticipating, and in most cases leading, the increasingly stringent legislative environment in which the industry operates worldwide.

As a primary business driver environmental considerations are to be seen in:-

(i) the re-emergence of gas as the most environmentally acceptable hydrocarbon

(ii) the downrating of 'environmentally difficult' oil development prospects/discoveries relative to more acceptable alternatives.

(iii) the intensifying search for additional oil within existing fields where the infrastructure and operating parameters are covered by known or predictable environmental standards. This activity is strongly linked to current exploration economics, of course, and therefore it would be wrong to attribute environmental considerations too strongly as a major driver. Nevertheless it is a contributing factor in the economic assessment of where investment should be placed.

2.3 Technological

2.3.1 Improved Oil Recovery Technologies

Industry perceptions of the improved oil recovery technologies which are deemed to be those *'most likely to succeed'* in the next five years are converging rapidly. With this convergence has come a clear prioritisation of where research and applications activity should be concentrated.

A recent review of oil company determined IOR targets (ref 4), by the UK DTI is summarised in Table 1. This is supported by the decline in publications citations on chemical injection and EOR (Figure 1).

Table 1: Perceived UKCS IOR Potential (After DTI)

Technique	Possible Oil MMBOE
-Horizontal/Extended Reach Drilling	2400
-Gas Injection	1425
-Depressurisation	800
-Flow Diversion	500
-Viscous Oil Recovery	200
-Modified Waterflood	40
TOTAL	5365

It is clear from Table 1 that the industry will concentrate on locating horizontal wells for best effect. This will be achieved by the current oil company push on 'Reservoir Definition Technologies', especially 4-D, or time-lapse 3-D seismic, well-to-well seismic, and other emerging geophysical technologies.

With the exception of deep diverting gels, and, possibly, the application of in-situ (microbial) generation of chemicals, the economic potential of chemically-enhanced recovery schemes is seen as unimpressive.

Does this realignment of effort mean the complete death of the use of flood-mode chemicals (as opposed to well treatments) in the reservoir? The answer in the long term is 'not necessarily', for improved reservoir description can help reduce uncertainty in the application of chemical treatments, improving the knowledge of more precise target applications. In the first instance, this should lead to more successful well treatments, and will play a large part in the take up of the deep gel technology which is now moving from experimental technique to available technology. The potential for the combination of deep gel plus horizontal well application has yet to be unlocked. Undoubtedly the time for the marriage of these two recent technologies will soon arrive as confidence builds in the efficacy of the chemical element.

2.3.1 Production Process and Drilling Systems

The currentenvironmental pressures on operators will continue to build. Where at possible, chemical usage will be reduced, then replaced with more environmentally acceptable alternatives as and when these become available. Ultimately improved knowledge, resulting from research on process operating parameters and materials may lead to the phasing out of chemicals which are currently used for 'insurance' purposes - e.g. hydrates and corrosion inhibitors.

On the drilling side the trend for the growing use of horizontal wells, will serve to reduce the number of development wells required, particularly offshore. Slimhole, or coiled-tubing drilling, currently under development, will also cut the mud volumes required in exploration and appraisal drilling, reducing usage, costs and adverse environmental loadings.

3.0 Market Activity Forecasts

Recent estimates by Scottish Enterprise (Ref 5) which were made at a projected oil price of $18/barrel suggest that the UKCS market for drilling mud additives and production chemicals will remain static, or slowly decline, in the range £80-90 million/year, for combined E&P activity. This figure does not include primary drilling mud components (baryte or base oil) or frac fluids.

On a similar basis, the equivalent worldwide market size is estimated to be approximately £800 million/year. A market need, expressed in terms of the current products which are sold within in the scope of a figure of this size, will not disappear overnight. But the share of the market which any one company will take will undoubtedly be influenced by the perceived environmental acceptability of the products on offer.

4.0 Response of Chemical Supply industry to business conditions

It is not particularly easy to characterise the supply industry. It is in no way 'homogeneous', containing small specialists to large multinational oil companies who have an interest in both the manufacture and supply side as well as the use and application.

What seems to be emerging, if the somewhat gross parallel can be drawn between the earth as an environmental entity, and the human body, is that the oilfield chemical industry will increasingly take on the features of the pharmaceutical industry.

The prescribing of chemicals will be avoided if at all possible, but the use will not be eliminated. Chemicals with demonstrated minimum side effects (on the environment) will prosper for a while until safer more effective replacements are researched and tested. Highly effective specialist molecules or cocktails should command a premium price whilst protected by patent. These patents will be achieved by high levels of research investment targeted on specialist problems, possibly developed in conjunction with universities who are leading the field in computer design techniques.

Other comparable features of the pharmaceutical industry might be made. Probably the most significant of these should be greater recognition of the oil industry equivalents of the prescribing physician - i.e. the production and mud chemists.   Dissemination of information to and within the industry will become more intensive; the growth and extension of current databases to include more comprehensive application data than currently exists is essential, and has specific parallels with medical applications.

5.0 Emerging Trends

There are a number of signs of where the oilfield chemicals industry is going Some companies are already part way along the way. Three potential avenues are signposted:-

(i) development and use of multifunction 'designer' molecules, which can replace more than one chemical currently being applied. Field specific design is a possibility. The closest reported development is the Miller field scale inhibitor. This was a for single purpose application. It is believed that development work on true multifunction chemicals is being advanced by more than one company. If such chemicals can be synthesised from naturally sourced base products then this will serve to minimise the ecological impact if such chemicals are discharged.

(ii) rapid growth in applications of bioscience to oil industry processes. The use of biotechnological intervention in other extractive industries is now routine. The oil industry has toyed with a number of applications. Given the current volume of research in bioscience, and likely industry spin-offs, the growth of a raft of new start-up companies with oil-related products, is a forecast for the 5-10 year horizon.

(iii) as a more 'off the wall' and uncertain prediction, but which nevertheless may be of interest to some, the oilfield chemistry industry might start to follow the current geoscience-based research which is starting to quantify the effect on reservoir behaviour of combined changes in reservoir stress state and chemistry. Why? Because emerging results show that measurable changes in rock characteristics (permeability/wettability etc) can happen in short timescales through the effect of diagenesis induced by changes in chemistry. This rock effect, probably best equated to stress corrosion cracking in steels, can happen under standard waterfloods. When better understood, the knowledge being produced on this phenomenon could re-open chemical flooding economics by being able to approach reservoirs with very cheap commodity products rather than exotic, high cost chemicals. If controlled management of chemical potential can favourably alter the rock properties, then potential opportunities, which offer real returns to oil companies may open up. It is early days - but understanding of the reservoir processes is building rapidly in a number of universities.

A lateral-thinking variant of this might possibly lead an adventurous chemical supply company to develop a new angle on reservoir characterisation, adding an alternative to tracer use. How? Because changes in chemical potential can lead to microfracturing, and this process releases energy as an acoustic signal. It may be beyond the capability of current instrumentation, and still in the realms of imagination, but use of a downhole geophone to measure the effect of chemical injection, could assist in decoding details of reservoir architecture in a truly ambitious interdisciplinary process.

6.0 Conclusions

The principal conclusions which can be drawn from this overview are as follows:-

(a) low prices will maintain the dominant position of hydrocarbons as a primary energy source. The 5-10 year world activity predictions imply a 'little or no growth' future for the oilfield chemicals market as a whole. Whilst drilling may remain depressed as oil companies pace exploration to cash flow, there will be focus and emphasis on maximising existing well productivities. Stimulation activity should stay firm.

For new production wells, companies who can offer a systems approach which spans chemicals application from drilling through completion and long term well performance will stand to gain under the partnership arrangements now sought by most oil companies.

(b) within the overall chemicals market, opportunities for most opportunities growth will come from substitution. Companies that can use brainpower to take the lead to develop or introduce more environmentally acceptable products and processes will be able to capitalise on their expertise. In this respect the oilfield chemicals industry is developing features characteristic of the pharmaceutical business. Investment in research will become a key parameter.

(c) oil industry-led improved oil recovery research is now concentrated on reservoir definition and use of horizontal wells. From the perspective of chemical supply, likely take up of deep diverting gel processes will see this technology -developed with persistance by BP and others- move rapidly from research to application. Take up is likely to climb, if not spectacularly, then steadily, in the next few years. Acceptance of this process may build confidence in use of other chemical processes

(d) wise oilfield chemical companies will be continuously screeing biotechnology opportunities for potential technology transfer to the oil industry.

(e) current industry-sponsored research into coupled chemistry and stress-state reservoir behaviour may create new knowledge which could be employed to the benefit of both oil companies and the oilfield chemicals industry in the medium to long term.

References

1. Charles M. Hudgins Jr, "Chemical use in North Sea Oil and Gas E&P". JPT jan 1994

2. Koos Visser, Shell Paper "The Test of Tomorrow", Sept 1993

3. BP Annual Reports, 1970,1971

4.B Coleman, DTI IOR conference "Best practices for Improved Oil Recovery" London Nov 1993

5.Scottish Enterprise, "Forecast of Upstream Petroleum- Activity and Expenditure UK and Worldwide 1993-1997"

Low Tension Polymer Flood. The Influence of Surfactant–Polymer Interaction

K. Taugbøl, He-Hua Zhou, and T. Austad

ROGALAND RESEARCH INSTITUTE, PO BOX 2503, ULLANDHAUG, N-4004
STAVANGER, NORWAY

ABSTRACT

Coinjection of low-concentration surfactant and a biopolymer, followed by a polymer
buffer for mobility control, leads to reduced chemical consumption and high oil
recovery. The method has been termed Low Tension Polymer Flood, LTPF, by BP.
The present paper gives a discussion about possible synergistic effects between the
surfactant and the polymer in a dynamic flood situation. Core flood experiments are
conducted using an aklylxylene sulfonate, xanthan, model oil (n-C_7), NaCl-brine, and
Berea sandstone cores in the two-phase region at 50°C. The chromatographic
separation of surfactant and polymer is very important to obtain good oil recovery and
low surfactant retention. At low surfactant concentration the flooding behavior of the
surfactant is influenced by the presence of polymer in a negative way regarding oil
recovery. The surfactant-polymer interaction has been discussed in terms of a weak
associative complex formation.

INTRODUCTION

After waterflooding of a sandstone oil reservoir considerable amounts of oil are
usually left behind. The residual saturation of oil after a waterflood is in the range of
$0.3 < S_{orw} < 0.5$. The oil is trapped in the pores due to capillary forces. The target for
micellar flooding is usually the waterflooded residual oil. In order to apply this type of
chemical improved oil recovery, IOR, the efficiency of the technique must be
significantly improved because of economical reasons. This means that more extra oil
must be produced at a lower input of chemicals.

Laboratory experiments conducted by Kalpakcy[1] et al. have shown that
coinjection of surfactant and polymer gives good oil recovery and low surfactant
retention. The method has been termed Low Tension Polymer Flood, LTPF. In a
recent paper we have discussed possible flooding mechanisms based on published
literature.[2] It was proposed that surfactant-polymer interaction will play an important

role in obtaining good results. Furthermore, a chromatographic separation of surfactant and polymer, i. e. polymer ahead of the surfactant, will probably decrease surfactant adsorption and may be improve the microscopic sweep efficiency.

In order to obtain optimal flood behavior, i. e. mobility control and low interfacial tension, the surfactant system is normally designed to give a three-phase region, and to use a salinity gradient to prevent dispersion and strong retention of the surfactant slug.[3] Sea water is the injection fluid in off-shore oil reservoirs, and the need to adjust the salinity at least ±10 % will increase the total cost. As an alternative, a polymer gradient was tested in core floods based on the fact that a decrease in the concentration of polymer will promote a phase transition from III to II(-).[4] This will control the phase behavior at the rear of the slug to avoid trapping of surfactant in the middle phase. It was concluded that no significant difference in the oil recovery was observed by comparing floods conducted by salinity and polymer gradients. However, due to an increase in mobility by decreasing the polymer concentration of the injection water, it is hard to obtain a sharp polymer gradient in the porous media at the rear of the surfactant slug.

Regarding North Sea oil reservoirs, the mobility ratio between the injection water (sea water) and the oil is favorable due to the low viscosity of the crude oil. This benefit must also be used in low concentration surfactant flooding in order to lower the costs of polymer for mobility control. It is, however, well known that the success of field tests is very much related to the amount of polymer that is injected behind the surfactant slug.[5] The surfactant concentration in the slug was usually high, about 5 wt%, and it was needed a rather viscous solution to push the surfactant slug through the formation. For low viscosity oils it is expected that injection of a low concentration surfactant slug, 0.1-0.5 wt%, in the II(-) state will need moderate amount of polymer to maintain mobility control.

The influence of the surfactant-polymer interaction, SPI, during a LTPF is probably rather sensitive to the phase behavior of the flooding system. In the three-phase state, III, the surfactant and the water soluble polymer is present in different phases, i.e. the surfactant will mainly stay in middle phase and the polymer is in the water phase. By conducting the flood using an oil-in-water microemulsion, II(-) state, both of the chemicals are in the water phase. In the latter case the flood behavior is simpler, and it may be looked upon as a low tension water flood.

The present paper describes some results from flooding experiments performed at a rather low surfactant concentration, 0.5 and 0.1 wt%, using an oil-in-water microemulsion, II(-). The surfactant is the same as used in the previous polymer gradient experiments, namely an alkyl xylene sulfonate. It is of special interest to study the effect of polymer, xanthan, in relation to oil recovery and flooding behavior of the surfactant.

EXPERIMENTAL

Material

Surfactant
The Exxon chemical termed RL-3011, dodecyl-ortho-xylene-sulfonate, was used. The material was purified as reported elsewhere.[6] NMR and HPLC studies showed that the surfactant probably contained different isomers. In all cases the surfactant was dissolved in 2.0 wt% NaCl solution.

Polymer
Xanthan was supplied from Statoil at a concentration of 2.9 wt%. The fermentation liquid was dissolved in a 0.2 wt% KCl-solution. Cell debrise and some aggregates were removed by centrifugation. Pure xanthan was precipitated by adding isopropanol. The xanthan was diluted to 4000 ppm in NaCl-brine containing 1000 ppm formaldehyde. Prior to use, the xanthan solution was filtered through Millipore filters of 5.0, 3.0, 1.2, and 0.8 µm.

Oil
n-heptane was used as model oil.

Brine
2.0 wt% NaCl dissolved in distilled water was used.

Cores
Standard Berea cores were used. The permeability is about 500 mD, and the length and the diameter is about 60 cm and 3.8 cm.

Flooding procedure

The 500 mD Berea cores were installed in Hasler core holders and placed in a water bath at 50 °C. Prior to the flood experiments, the cores were flooded with 4 PV of brine. Four surfactant floods have been performed with the following fluid injection sequence:

Flood 1:   0.5 PV of 0.5 wt% surfactant followed by brine.
Flood 2:   0.5 PV of 0.5 wt% surfactant and 500 ppm xanthan, 1.0 PV of 250 ppm xanthan followed by brine.
Flood 3:   0.5 PV of 0.1 wt% surfactant followed by brine.
Flood 4:   0.5 PV of 0.1 wt% surfactant and 500 ppm xanthan, 1.0 PV of 250 ppm xanthan followed by brine.

The flow rate was kept constant at 0.4 ml/min in all cases. The differential pressure over the cores was determined during the flooding period. It was carefully verified that the system showed an oil-in-water microemulsion, II(-) state, at the flooding conditions. Furthermore, it was verified that the chemicals were compatible at the present concentrations and temperature, i. e. no phase separation or precipitation took place.

Interfacial tension

The interfacial tension, IFT, between the oil and the brine, with and without xanthan present, was determined using a spinning drop tensiometer.

Chemical analysis

The concentration of surfactant was determined by two-phase tritration according to Reid et al.[7] The concentration of xanthan was determined using size exclusion chromatography, GPC. The column was of the type Waters Ultrahydrogel 250. Brine was used as the mobile phase at a flow rate of 0.7 ml/min. The polymer was detected using a refractive index detector. Calibration runs showed a linear relationship between detector response and xanthan concentration in the range of 0-500 ppm. The chromatographic analysis were performed on a HPLC system delivered by Waters.

RESULTS AND DISCUSSION

Physical data of the respective cores, residual oil saturations and end-point permeabilities, and flooding efficiency and retentions for the various floods are summarized in Table 1 a, b, and c, respectively. It is seen that the absolute permeability varies between 530-620 mD. The irreducible water saturation, S_{wi}, is about 0.30 and the oil saturation of the waterflooded cores, S_{orw}, is about 0.34. The final oil saturation after the chemical floods, S_{orc}, ranged between 0.08 and 0.28.

In all cases the cores were waterflooded to obtain S_{orw} prior to the injection of chemicals. The extra oil recovery from Floods 1 and 2, using 0.5 wt% surfactant, is presented in Fig. 1. It is of interest to note that in both cases oil is produced after injecting less than 0.2 PV of the surfactant solution. According to the oil production profile of Fig. 1, the displacement of oil appears to be quite stable. The oil cut at the core outlet stabilized between 20 and 30%. It is, however, a surprise that the oil recovery is significantly greater without using polymer in the injected surfactant solution. The oil recovered relative to S_{orw} is 76 and 62% for Flood 1 and 2, respectively. This is also reflected in the value of S_{orc} which is 0.08 and 0.13 for the two floods. From Table 1c it is seen that the flooding efficiency is about three times higher, and the surfactant retention is also lower in the case without xanthan present.

A similar tendency regarding extra oil recovery is observed for Floods 3 and 4, using only 0.1 wt% surfactant. The relative decrease in oil recovery in the presence of xanthan is more pronounced in the low surfactant concentration case. Furthermore, the shape of the oil production profiles in Fig. 2 suggests a stable displacement for the system without xanthan present. In the presence of xanthan the shape of the curve below 1 PV is rather curved and it is not in agreement with a stable displacement. As observed from Table 1b, the values of S_{orc}, 0.207 and 0.276 for Floods 3 and 4 respectively, are considerably higher than in Floods 1 and 2. However, due to the very small amount of surfactant injected, the efficiency of the floods is very high, 260 and 138 for Floods 3 and 4, respectively. Thus, these experiments at least show that it should be possible to design micellar flooding of oil reservoirs in a cost effective way even in the II(-) state.

Table 1.

a. Physical data of the cores

F l	Length (cm)	Diameter (cm)	Weight (g)	PV (cm^3)	Porosity	Ab.Perm. (mD)	Rate (ml/min)
1	60.0	3.78	1357	149.0	0.22	560	0.40
2	59.3	3.76		133.6	0.203	619	0.40
3	59.45	3.77	1344	148.4	0.22	593	0.40
4	59.4	3.78	1342	148.0	0.22	529	0.40

b. Residual saturations and end point permeabilities.

Flood	S_{wi}	S_{orw}	S_{orc}	k_{ro} (S_{wi})	k_{rw} (S_{orw})	k_{rw} (S_{orc})	Surf. con. (wt%)
1	0.308	0.329	0.080	0.988	0.105	0.720	0.50
2	0.261	0.347	0.130	0.718	0.085	0.56	0.50
3	0.312	0.337	0.207	0.777	0.066	0.111	0.10
4	0.303	0.345	0.276	0.924	0.079	0.104	0.10

c. Efficiency and retention

Flood	Efficiency	Surf. ret. (mg/g)
1	100	0.15
2	29	0.23
3	260	0.042
4	138	0.037

Efficiency: ml oil / g surfactant injected.
Surfactant retention: mg / g reservoir rock

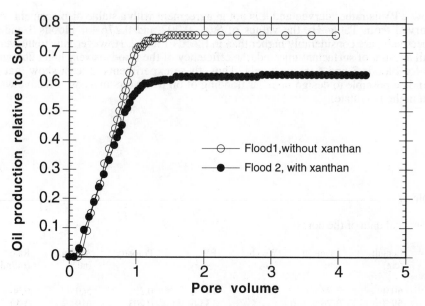

Fig. 1. Cumulative oil production for Flood 1 and 2 using 0.5 wt% surfactant slug.

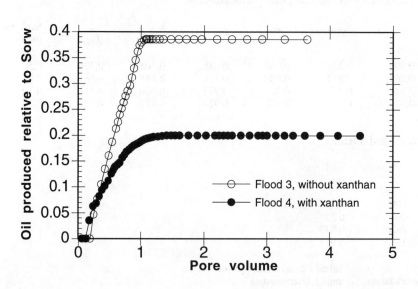

Fig.2. Cumulative oil production for Flood 3 and 4 using 0.1 wt% surfactant slug.

The general knowledge obtained from the results so fare is that the polymer decreases the efficiency of the floods at both surfactant concentrations. This is not in accordance with what should be expected from a LTPF. Several mechanisms can influence the flooding behavior of the surfactant slug resulting in the observed phenomena:

a.   The polymer may affect the mobility ratio between the injected and the displaced fluid in a bad direction.

b.   The polymer can increase the IFT between oil and brine by associative interaction with the surfactant.

c.   The polymer and the surfactant can make associative interaction in such a way that the surfactant will mainly stick to the polymer during the flooding process.

According to Table 1b, the relative permeability of water, k_{rw}, increases dramatically upon increasing the water saturation from S_{orw} to S_{orc}. The viscosity of brine and n-heptane at 50°C is 0.56 and 0.32 cP, respectively. The viscosity of 500 and 250 ppm xanthan solution in the porous media is determined to be 2.7-3.4 cP and 1.3-1.6 cP, respectively. Thus, 500 ppm xanthan will decrease the mobility relative to brine by a factor 4.8-6.0. In the case of 250 ppm xanthan the factor is 2.3-2.8. From a mobility point of view the presence of polymer should prevent dispersion of the surfactant slug and improve the oil recovery. Chromatographic separation of surfactant and polymer will take place during the flooding process which can disturb the mobility conditions. This will be discussed later.

The IFT between brine and oil versus the surfactant concentration, with and without xanthan, is shown in Fig. 3. It appears that xanthan at 500 ppm does not affect the IFT significantly. Thus, the decrease in oil recovery by adding xanthan can not be related to increase in IFT. The similar shape of the two curves also indicates that strong associative interaction between the surfactant and the polymer does not exists.[2] Furthermore, according to Figure 3 the critical micelle concentration, CMC, of the surfactant in the brine at 50°C is at about 0.01 wt%. Thus, the 0.1 wt% surfactant system can only tolerate a ten time dilution in order to keep the low IFT. Kalpakcy[1] and coworkers noticed a decrease in IFT when adding polymer to a surfactant formulation suitable for LTPF, which is not observed in the present case.

The surfactant elution profile for Floods 1 and 2 and Floods 3 and 4 is shown by Figs. 4 and 5, respectively. The surfactant effluent profile for the 0.5 wt% surfactant system, Fig. 4, is completely different when xanthan is present. Fist, the elution of surfactant starts at a significantly lower PV, in fact well below 1 PV. Second, no clear peak in the surfactant concentration was detected, which indicates a strong dispersion of the surfactant slug. The average concentration of surfactant is about 2.5 % of the injected concentration, which corresponds to about 0.013 wt%. This is slightly above the CMC which means that the low IFT is maintained during the flood. It is of interest to note that in the interval 0-2 PV 35.8% of the injected surfactant was eluted for Flood 1, while only 4.8 % was eluted for the xanthan containing system. The very sharp increase in surfactant concentration at 1 PV for Flood 1 and the rather symmetric surfactant peak further strengthen the suggestion of a stable displacement.

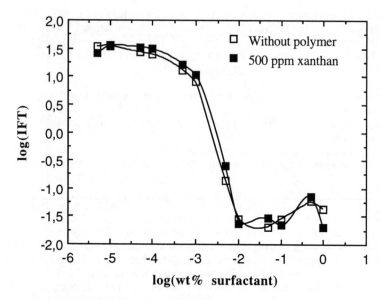

Fig. 3. Interfacial tension, IFT, between n-C7 and brine at 50 °C.

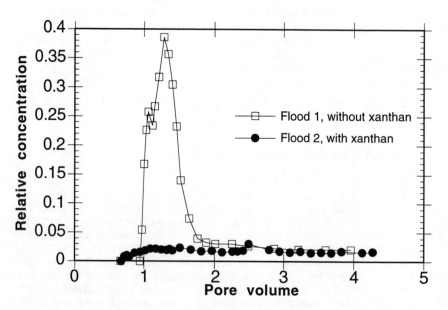

Fig. 4. Elution profile of the surfactant for Floods 1 and 2 using 0.5 wt% slug.

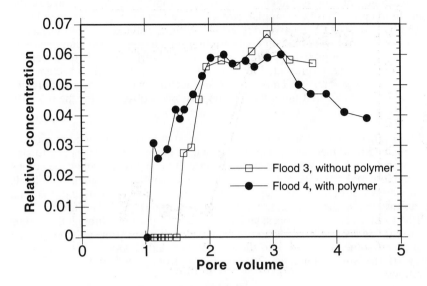

Fig. 5. Elution profile of the surfactant for Flood 3 and 4 using 0.1 wt% slug.

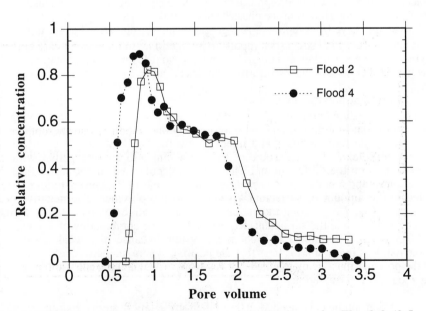

Fig. 6. Effluent profile of the polymer vs. pore volume for Flood 2 (0.5 wt% surfactant) and Flood 4 (0.1 wt% surfactant).

Concerning the low surfactant concentration floods, Fig. 5, the shape of the surfactant elution profiles is quite similar. However, the surfactant production in Flood 3 is delayed by about 0.5 PV relative to Flood 4 containing xanthan. Due to surfactant adsorption and a very low surfactant concentration the elution of surfactant starts at about 1.5 PV.In the interval 0-2 PV 3.6 % of the injected surfactant was produced for Flood 3. When xanthan is present, Flood 4, 7.4 % of the injected surfactant was produced in the same interval. Thus, at low surfactant concentration the polymer appears to enhance surfactant mobility. The opposite was observed for the 0.5 wt% surfactant flood.

According to the discussion above, it appears that the experimental results can be best explained by a mechanism described by c. It is well known that due to size exclusion or inaccessible pore volume the polymer will move faster through the porous medium than low molecular weight species. The effluent profiles of xanthan in Floods 2 and 4 are shown in Fig. 6. The shape of the profiles is quite similar, and it is seen that it is a sharp rise in the polymer concentration just after 0.5 PV. This means that it is established water-continous zones, which transport the polymer through the porous medium, excluding a significant fraction of the pores. It appears as the polymer has the ability to divert some surfactant into the same zones and in this way lower sweep efficiency of the surfactant slug. This is more pronounced at low surfactant concentrations.

It is well known in the surface chemistry that surfactant and polymer can make associative complexes.[2,8,9] Mixtures of nonionic polymers and ionic surfactants and surfactant-polymer mixtures of opposite charge have been studied most extensively.[8,9] Very few systematic studies of the phase behavior of similarly charged polyelectrolytes and surfactants have been reported. A segregative phase separation has been reported in the latter case which suggests a repulsive interaction.[9] In the present case both the polymer and the surfactant is negatively charged, and from an electrostatic point of view it is hard to believe that associative complexes can be formed.

Surfactant-polymer interaction can be studied by means of "dynamic dialysis" using gel permeation chromatography, GPC.[10] The test is conducted by dissolving the surfactant in the mobile phase at an appropriate concentration, and the polymer is dissolved in the mobile phase and injected into the GPC system. If a surfactant-polymer complex is formed, some of the surfactant will move together with the polymer through the GPC column. The large molecules will move faster than the smaller ones, and a negative or vacant peak will appear in the chromatogram that is related to the amount of surfactant associated to the polymer. The negative peak observed in Fig. 7 is a strong indication that an associative complex can be formed between xanthan and the surfactant RL-3011. Thus, the combination of xanthan and RL-3011 is not a suitable LTPF chemical system because of a weak associative interaction. It is important to stress that the positive synergism between the surfactant and the polymer must be maintained at a very low surfactant concentration in order to be a good LTPF formulation.

Further studies are needed, using other surfactant systems that do not make complexes with the polymer, in order to understand the simultaneous flow of polymer and surfactant in porous media at low concentration. It is well known that ethoxylated

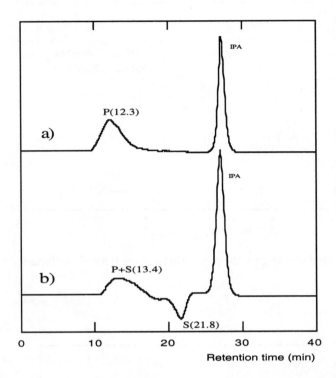

Fig. 7. GPC-analysis confirming that a complex between RL-3011 and xanthan is formed. (a) Without surfactant in the mobile phase. (b) With surfactant in the mobile phase.[10]

sulfonates/sulfates have a less tendency to form complexes with polymers.[11] Further studies are in progress at our laboratory.

The differential pressure over the cores is pictured in Figs. 8 and 9. Without polymer present the differential pressure decreases during the oil production period, and then it is stabilized. In the presence of xanthan the differential pressure increases to a maximum. After about 2 PV , when the polymer has been eluted, the differential pressure is quite similar to the systems without polymer. Thus, the pressure behavior is quite normal confirming good flow behavior.

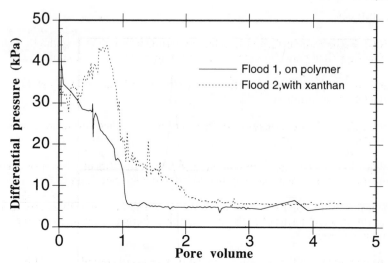

Fig. 8. Differential pressure vs. the pore volume using 0.5 wt% surfactant.

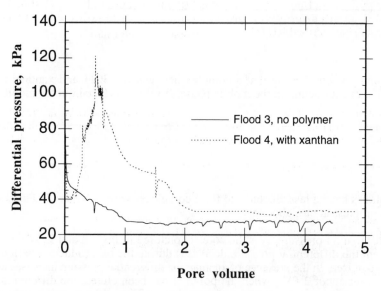

Fig. 9. Differential pressure vs. pore volume, using 0.1 wt% surfactant

CONCLUSIONS

Base on the four surfactant flooding experiments conducted in the II(-) state at a rather low surfactant concentration, some preliminary statements can me made:

Xanthan, 500 ppm, does not affect the IFT between brine and n-C_7 at different surfactant concentrations.

Very high efficiency of residual oil recovery was obtained surfactant floods in the II(-) state.

Coinjection of surfactant and polymer resulted in significantly lower oil recovery compared to floods without polymer added to the injected fluid.

The influence of polymer on the oil recovery and surfactant effluent profile has been discussed in terms of associative complex formation between the surfactant and the polymer.

It is concluded that the present chemical system is not suitable for LTPF.

ACKNOWLEDGEMENT

The project is funded by the state and industry (Statoil, Norsk Hydro, Total and Saga) supported program on Reservoir Utilization through advanced Technological Help, termed RUTH.

REFERENCES

1.   B. Kalpakcy, T. G. Arf, J. W. Barker, A. S. Krupa, J. C. Morgan and R. D. Neira, "The Low-Tension Polymer Flood Approach to Cost-Effective Chemical EOR", Paper SPE/DOE 20220, presented at the 7th. Symposium on Enhanced Oil Recovery, Tulsa, Ok.,April 22-25,1990.

2.   T. Austad, I. Fjelde, K. Veggeland and K. Taugbøl, J. Pet. Sci. Eng., 1993, in press.

3.   C. Nelson, Soc. Pet. Eng. AIME, 1982, 227-238.

4.   T. Austad and K. Taugbøl, "Polymer gradient as an alternative to the salinity gradient for controlling the effects of dispersion and retention in LTPF", paper presented at the 4th. IEA Collaborative Project on Enhanced Oil Recovery, Salzburg, October 17-21, 1993.

5.   L. W. Lake and G. A. Pope, Petr. Eng. Int., 1978, 51, 38-60.

6.   T. Austad and G. Staurland, In Situ, 1990, 14 (4), 429-454.

7.   V. W. Reid, G. F. Longman and E. Heinerth, Tensid, 1967, 4, 292-304.

8.   E. D. Goddard, <u>Colloids and Surfaces</u>, 1986, <u>19</u>, 255-300.

9.   L. Piculell and B. Lindman, <u>Advances in Colloid Interface Sci.</u>, 1992, <u>41</u>,149-178.

10. K. Veggeland and T. Austad, <u>Colloids and Surfaces A</u>, 1993, <u>76</u>, 73-80.

11. S. Saito, <u>J. Colloid Interface Sci.</u>, 1960, <u>15</u>, 283-286.

Chemical Gel Systems for Improved Oil Recovery

H. Frampton

ALLIED COLLOIDS LIMITED, PO BOX 38, LOW MOOR, BRADFORD, WEST YORKSHIRE BD12 0JZ, UK

1. Introduction.

Poor distribution of injected fluids and unacceptably high levels of water or gas production are common and costly problems in oilfields worldwide. In particular the presence of high permeability thief zones which cause rapid waterflood communication from injector to producer is common in the oilfields of the northern North Sea. These problems can be treated using chemical systems which form gels to divert fluid flow. The intention is to direct a relatively low viscosity mixture of reagents into the path of the fluid where it sets into a gel and diverts the flow of the fluid allowing more effective oil production.

Thousands of gel treatments, of numerous types, have been applied over the last two decades. Improvements in success rate from an initial 30% to a current 50 - 70% have been achieved through developments in chemistry and application technology. This has resulted in treatments which are at least as reliable as squeeze cementing which is considered a routine well workover technique. With such success and field experience it is surprising that gel treatments are not regarded as an everyday tool of the reservoir engineer. The fact that they are not can perhaps be attributed to the following reasons (amongst others)

1. Too many chemical options have been published without being commercially available.
2. Chemical gels are perceived to be exotic and potentially unreliable.

3. The cost and risk of treatments are perceived to
    outweigh the potential reward.
This paper sets out to compare, in a practical way, four
commercial gel systems which have been used in North Sea
wells.   The aim is to present facts which will allow a
clearer   understanding   of   the   chemistry   behind   the
application.
          2. **The history of chemical gel systems for fluid
             diversion.**
The concept of using chemical gel systems for fluid
diversion in oil and gas producing formations seem to
have originated in the mid-1960's.   Table A gives a
selective chronological summary of the main commercial
systems which have achieved field use.

Four systems stand out for use in the North sea.   Drawn
from the larger list of systems in use these are ;
1. Low molecular weight polyacrylamide copolymer
    crosslinked with chromium ions. (PACM/CrAc)
2. Xanthan Gum crosslinked with chromium ions.(XG/CrAc)
3. Poly(Vinyl alcohol) crosslinked with
    glutaraldehyde.(PVOH/Glut)
4. Acidified sodium silicate. (Silicate)
All of these have been used in North sea well treatments
within the last three years.

          TABLE A.
A selective chronological summary of the development of
water shut-off gel systems.

APPROXIMATE
YEAR   **SYSTEM**

1965   Monomer gels
1967   Polyacrylamide + Polyvalent metal ions
1973   Polyacrylamide + Chromium(VI) reduction
1973   Xanthan + Chromium(VI) reduction
1974   Polyacrylamide + Aluminium Citrate
1977   Silicate
1979   Polyacrylamide + Aldehydes
1983   Lignosulphonate + Chromium(VI) reduction
1985   Poly(Vinyl alcohol) + Glutaraldehyde
1985   Melamine Formaldehyde
1987   Polyacrylamide + Phenol Formaldehyde
1987   Polyacrylamide + Chromium(III) Chelates
1987   Xanthan + Chromium(III) Ions

          3. **Crosslinking mechanisms.**
3.1 Polyacrylamide copolymers with chromium acetate.
The   term   "Polyacrylamide   copolymers"   traditionally

encompasses all copolymers of acrylamide and sodium acrylate. Only copolymers with a low anionic content (weight percent sodium acrylate) and low molecular weight currently appear suitable for use under North sea conditions. Such products can be supplied in various physical forms; the most appropriate for any given application depends on the size of the treatment, the handling facilities available and other logistical considerations. In general the products are easily handled and dissolved.

Chromium acetate is supplied as a low viscosity, dark green, 50% active solution. The product is typically of pH about 3 and smells slightly of acetic acid. Unlike chromium (VI) complexes, Chromium (III) complexes exhibit low acute toxicity (1,2,3).

Chromium ions in carboxylate complexes are in equilibrium with free ions in solution. Using Chromium Acetate as an example:

1. $\quad [Cr(CH_3COO)_3(H_2O)_3] \rightarrow [Cr(CH_3COO)_2(H_2O)_4]^+$
$\quad\quad\quad\quad\quad\quad\quad\quad\quad\quad\quad\quad\quad\quad\quad\quad + CH_3COO^-$
$\quad\quad\quad\quad\quad\quad\quad\quad\quad\quad\quad\quad\quad\quad\quad\quad\quad\quad\quad \downarrow$

$\quad [Cr(H_2O)_6]^{3+} \quad\quad\quad \leftarrow \quad [Cr(CH_3COO)(H_2O)_5]^{2+}$

$\quad + CH_3COO- \quad\quad\quad\quad + CH_3COO^-$

Chromium (III) has been shown to crosslink between carbohydrate residues in the form of a binuclear complex (6). In the crosslinking reaction, the displacement of the ligand on the binuclear chromium complex is a very slow step and is probably the rate limiting reaction in the gelation process (21). The chromium species involved in the gel formation and particularly the hydrated chromium complexes are susceptible to reaction with bidentate chelating groups which can be used to control the gel time at high temperatures (6).

About 3 to 5 weight percent of polymer is combined with 0.1 to 0.2 weight percent of chromium acetate to form gels. The gel strength required dictates the amount of polymer needed. Gel formulation is simply a matter of mixing the crosslinker and any retarder into the polymer solution.

3.2 Xanthan Gum with Chromium ions.

Xanthan is a natural heteropolysaccharide produced by fermentation of carbohydrates using the bacterium *Xanthomonas campestris* (8). The structure of a typical Xanthan molecule is shown in Figure 1. Minor variations on this are produced by mutant bacterial strains. Xanthan is usually supplied as a high viscosity (100,000 cP) concentrate containing up to 12.5 weight percent active

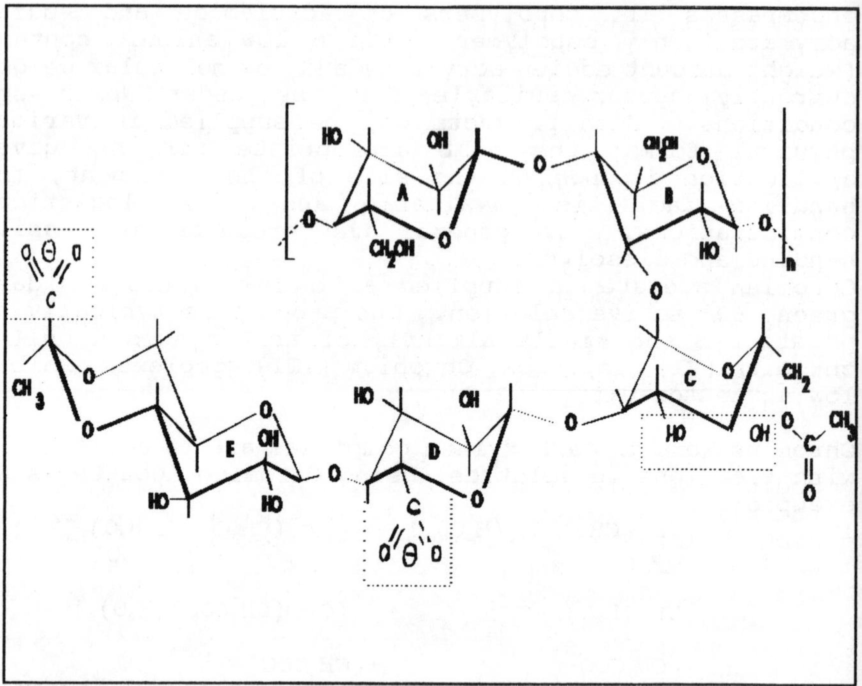

Figure 1 The molecular structure of Xanthan (13)

polymer. Special arrangements are needed to pump and
dilute it (9).
Xanthan Gum is not considered to be toxic even though it
contains   low   levels   of   preservatives,   usually
Formaldehyde (Methanal).

The mechanism for the crosslinking of Xanthan is similar
to that for Polyacrylamide copolymers.   The reactive
sites for the chromium ions are the carboxyl groups and
the cis - hydroxyl groups (12). These are highlighted in
Figure 1.   About 0.05 to 1.0 weight percent Xanthan is
combined in solution with 0.04 to 0.1 weight percent
chromium acetate to produce gels (11,12).   The high
viscosity of the gel mixture limits the maximum
concentration of polymer to below about 1.5% active.

3.3 Poly(Vinyl Alcohol) crosslinked with glutaraldehyde.
Poly(Vinyl Alcohol) [PVOH] polymers are manufactured from
Ethene Ethanoate (Vinyl Acetate) which is homopolymerised
and then hydrolysed to give Vinyl Alcohol co- Vinyl
Acetate.   The molecular weight and degree of hydrolysis
of the polymers are the major variables. PVOH polymers

are usually supplied as dry powders though concentrated (typically 8%) stock solutions can be prepared. The powders can be of relatively low bulk density. The dissolution of the polymers in cold water is very slow so heating to between 70 and 90°C is required to achieve acceptable rates. There is a strong tendency for the solutions to foam. A 12.5% w/v solution of 115,000 molecular weight PVOH in fresh water typically has a viscosity of about 100 cP.

Gels are prepared in brines of 10% Total Dissolved solids or below, by combining 2.5 weight percent of PVOH with 0.03 weight percent of Pentanedial (Glutaraldehyde) and adjusting the pH of the mixture to 2.7 or thereabouts with an acid such as methanoic acid (glacial acetic acid) (14).

3.4 Silica gels.

Sodium silicate is a complex mixture of species based on the silicate anion, SiO_2^-. It is typically supplied as a high pH solution with composition specified either by the active content and ratio of SiO_2 : Na_2O, or by the individual contents of SiO_2 and Na_2O. Solid forms are available but these tend to be difficult to dissolve. Commercial material used in the North Sea was recently specified as having a $SiO_2:Na_2O$ ratio of 3.3 : 1. A product with a $SiO_2:Na_2O$ ratio of 3:2 w/v is also available and was used in the laboratory work presented here. The dry weight (120°C, 24hours) of this product as supplied was measured as 44 percent w/v. The Brookfield UL viscosity versus shear rate

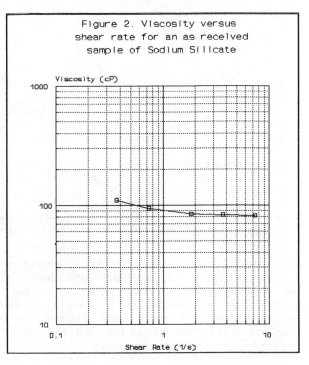

Figure 2. Viscosity versus shear rate for an as received sample of Sodium Silicate

profile is shown in Figure 2. Partial neutralisation of

dilute solutions of Sodium Silicate, for example with Hydrochloric acid, produces gels. The range of conditions under which gels are formed is relatively limited. For example, gels are formed from dilute solutions of sodium silicate at pH 8 to 9.
If the concentration of the resulting sodium ions is above 0.3M (15).
The first step in the mechanism of the formation of silica gels by neutralising sodium silicate solutions is the liberation of monosilicic acid. This is followed by the formation of dimers, trimers and higher polymers with linear and branched structures (16,17).

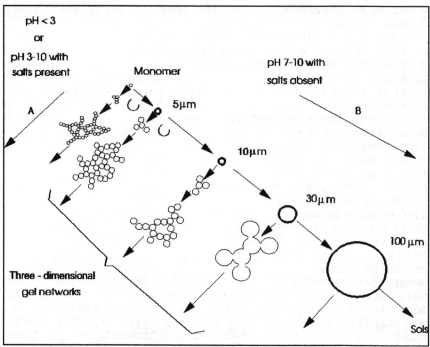

Figure 3   The polymerisation of silica (17)

In basic solution these rapidly undergo further condensation polymerisation and rearrange to form spherical silica particles. The mechanism can be summarised as shown in Figure 3 after Iler (17).
Water shut off gels are formed by neutralising 6%w/w sodium silicate solution to pH 10.2 to 11.0 with, for example, Hydrochloric acid.

4. The application of a chemical gel system.
Consideration of each stage in the supply, handling and use of a chemical gel system leads to the list of important properties of gel systems shown in Table B. It should not be inferred that this list is exhaustive.

Table B. Properties important in the application of chemical gel systems.

	Property	Desired value
1.	Number of components	Low
2.	Toxicity of any component	Low
3.	As - supplied concentration of any component	High
4.	Gel recipe concentrations	Low
5.	Cost per cubic metre of gel	Low
6.	Ease of mixing and formulation (Including equipment costs)	Easy
7.	Shear stability	High
8.	Injectivity	High
9.	Adsorption	Medium
10.	Pore - throat retention	Low
11.	Temperature sensitivity of gelation rate	High
12.	Gel salinity sensitivity	Low
13.	Gel pH sensitivity	Low
14.	Gel sensitivity to dilution	Low
15.	Gel strength	High
16.	Gel stability	High
17.	Degree of permeability reduction	High

4.1 Logistics.
Items 1 to 4 represent the logistics and handling of the gel system. Such considerations are often left to last when a gel system is selected, but they are the first and most obvious aspect of the actual use. For this reason it is important to include a consideration of these aspects in any screening programme. Table C compares the logistical properties of the four gel systems highlighted here. In part 3, A refers to the gel base, B the crosslinker and C the catalyst or retarder. Unbracketed values are the most common. The proportional price estimate is given for completeness, but is very variable. The figure given is based on the concentrations and recipes given, combined with quoted prices for bulk supply of the reagents. It does not include transport and application costs. The chemical cost was then proportionated to the cheapest chemical system (Silicate) to give the tabulated values.

Table C. Logistics of gel systems.

Property	PACM/CrAc	XG/CrAc	PVOH/Glut	Silicate
1. Number of components				
	3	3	3	2
2. Toxicity	Low	Low	Glut : **Harmful** Acetic Acid **Flammable & Corrosive**	Silicate : **Harmful** HCl : **Corrosive**

3. As supplied Concentrations (%)

A.	(12.5)-95	12.5-(95)	(12.5)-50	44
B.	50	50	50	15 - 30
C.	70	70	80 - 100	

4. Typical Gel Recipe concentrations (% active)

A.	4	0.9	2.5	6
B.	0.1	0.3	0.03	2
C.	2	2	0.5	

5. Proportional chemical cost per m³ Gel

	4.0	5.0	1.2	1.0

6. Ease of Mixing etc.

	Easy	Easy	Easy	Easy from solution

4.2 Injection.
Items 7 and 8 in Table B represent important considerations when the gel mixture is being pumped down the well and into the target zone. Shear stability during injection is not a concern at normal gel treatment injection rates of 1 to 3 Barrels per minute (0.16 - 0.48 m³/min) though in principle PACM and PVOH are vulnerable. All of the "North Sea" gel systems are composed of low molecular weight species and as such the injectivity of each is excellent and primarily controlled by the viscosity of the gel mixture. Typical viscosities are shown in Table D.

Table D. Typical Viscosities of Gel mixtures

PACM/CrAc	10 - 100 cP
Xanthan/CrAc	1000 - 10,000 cP
PVOH/Glutaraldehyde	1 - 20 cP
Silicate	1 - 2 cP

4.3 Adsorption.
Adsorption of reagents by reservoir rock takes place predominantly at the leading edge of the chemical slug.

This can lead to differential depletion of reagents. The nett result can be that the front edge of the chemical slug does not form a gel. This is most relevant for large slugs such as are needed for in - depth treatments. The adsorbed reagents are concentrated at the surface of the rock pore and since the gel time for most systems is inversely proportional to concentration this may lead to the rapid formation of gel at the pore walls. Adsorption is also a factor influencing the adhesion of gel. On Aluminosilicate glass and sandstone substrates polyacrylamide polymers adsorb more strongly than polysaccharide copolymers such as Xanthan. PACM/CrAc gels usually show strong adhesion to glass but XG/CrAc gels show less. The adsorption from high pH non - gelling silicate solutions onto mixed quartz/kaolinite /carbonate packs linearly increases with silicate concentration up to 0.06% and is in the region of 1.5 - 2.5 mg/cm³ at 2% silicate (19). The relative strengths of adsorption are summarised below.

Table E. Relative strengths of adsorption.

Order	PACM	>	PVOH	>	Silicate?	>	XG
Adsorption (μg/g)	40-85		-		-		2-55
(lb/acre ft)	250-500		-		-		10-330

Data from reference 20.

4.4 Pore throat Retention.
Retention of some or all of the active components of the gel mixture can be a dominant effect in the propagation and effectiveness of any gel system. Retention of reagents, as opposed to adsorption is unimportant for any low molecular weight components of any gel systems.
Of prime importance is the retention of gel during formation. Evidence suggests that any gel system must pass from monomeric species through oligomeric and will only form a bulk gel in the last stages of reaction. The implication is that gel aggregates form before gelation. These can be easily retained in pore throats (24,25,26). The time at which gel aggregates block pore throats can be regarded as a Gel time. It is usually shorter than the bulk gel time in a bottle and is more relevant for systems to be placed in matrix rock.

4.5 Gel sensitivity.
It is relatively important that a commercial gel system is chemically robust as there is always a degree of uncertainty in any application. This means that it should ideally be insensitive to most of the variables listed as 11 to 14 in Table B. The exception is temperature where it is advantageous if the gel time is

very long at surface temperatures, relative to the rate at reservoir temperature. The temperature sensitivity of most gel systems can be represented in terms of the Arrhenius equation (21). Figure 4. shows a comparison of gel systems where the data on silicate were drawn from the literature (20).

Gels formed from polymers with chromium acetate are relatively insensitive to salinity or brine composition (5). PVOH/Glut is also claimed to be (14) but in

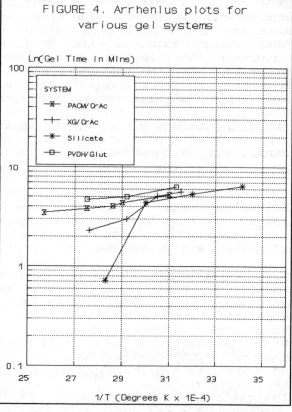

FIGURE 4. Arrhenius plots for various gel systems

Ln(Gel Time in Mins)

SYSTEM
—✕— PACM/CrAc
—+— XG/CrAc
—✱— Silicate
—☐— PVOH/Glut

1/T (Degrees K x 1E-4)

practice is limited in use to salinities below 10% (10,000 ppm) (20). This means that a fresh water supply is required. Silicate gels can only be formed adequately in fresh water because the high pH of the silicate precipitates the brine salts.

Sensitivity to dilution can be inferred from Figure 5. All four gel systems are similar in this respect.

pH sensitivity of the gel mixtures varies widely. PACM/CrAc gels have been reported to be effective over a wide range of pH (3.3 to 12.5) though the gel time changes significantly within this region (5). XG/CrAc gels can be inferred to have similar behaviour. Adjustment of pH with added acid is used to control the gel time of PVOH/Glut gels. The rate of gelation increases with pH decrease below about 5 and the lower limit of pH is about 2 (14). The silicate gel system is the most sensitive to pH. The gel time can vary from 1 hour at pH = 10.15 to 200 hours at pH = 11.0 (21). At pH's below 10 gels or precipitates form very rapidly. Homogeneous adjustment and measurement of pH to such accuracy is difficult. A further problem is the fact that the desirable pH lies so far above the normal reservoir pH. This gives rise to concern over premature gelation due to the buffering effect of the rock (19). A summary of the sensitivity of the gel systems to the four parameters highlighted is presented in Table F.

Table F. Summary of gel system sensitivity.

Sensitivity to :	Desired Value :	PACM/CrAc	XG/CrAc	PVOH/Glut	Silicate
11. Temp.	High	High	High	High	High
12. Salinity	Low	Low	Low	Medium	High
13. pH	Low	Low	Low	High	Very high
14. Dilution	Low	Medium	Medium	Medium	Medium

4.6 Gel strength.

Oscillatory rheology is commonly used to measure the strength of gels (28). It has been used to follow the kinetics of gelation (7). Two moduli are derived from experimental measurements:

G' - The storage modulus which may be thought of as a measure of "rubber-like" strength.

G'' - The loss modulus which may be regarded as a measure of viscous response.

The ratio G''/G' (termed Tan delta) is interpreted as a measure of the gel strength. The smaller the value, the more elastic the gel. Figures 5 and 6 compare the systems of concern in this paper.

Silicate systems give high values of G' but are very brittle. Polyacrylamide/Chromium acetate gels which give lower G' strengths are significantly more robust, some evidence of this can be seen in Figure 6. Other techniques, such as penetrometer measurements or yield pressure in a porous medium, have been used to obtain such information.

4.7 Gel stability.

There are two aspects of gel stability.
1. Syneresis (27,28)
2 Thermal degradation of the bulk gel.
In each case a gel formulation can be obtained with a good resistance to syneresis and minimal thermal degradation. In ageing tests on samples used in this work the gels formed with Chromium Acetate were least

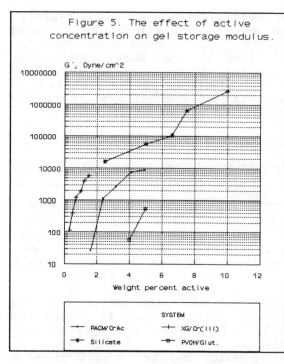

Figure 5. The effect of active concentration on gel storage modulus.

susceptible to syneresis. PVOH/Glut were more vulnerable and silicate gels were very prone to it. Polyacrylamide, Xanthan and Poly(Vinyl Alcohol) are known to be susceptible to free radical degradation in solution and gels can potentially break down over time at high temperature, however stability above and beyond that of the free polymer has been noted for gels (14,15).

4.8 Degree of permeability reduction.
The aqueous phase RRF range for mature gels was reviewed by Woods (20) for three of the four systems of interest here. Table G summarises the results. The values for silicate gels are estimated from the gel properties observed in the present work. It is interesting to note that the polymer based gels appear to selectively reduce water relative permeability rather than oil, but the silicate system does not (10).

Table G. Estimated ranges of aqueous phase RRF.

PACM/CrAc	XG/CrAc	PVOH/Glut	Silicate
10 − ∞	10 − ∞	≈ 10	10 − ∞

5. THE FUTURE OF WATER SHUT − OFF TREATMENTS.
As long as the relevant factors of any application are considered and allowed for, the setting of a chemical gel system has a pleasing inevitability. In many cases the success or failure of a treatment rests upon the ability to place the gel where it will perform its appointed task (29). To some extent the ability to achieve selectivity depends upon how well the target zone can be

characterised. Accordingly there is an increasing demand for systems which can be injected into the reservoir indiscriminately. A possible complication is that bulk gels all work in the region near the production well, for example a typical well separation in the North Sea is 1000 metres and a typical gel treatment radius is 5 to 15 metres. If no substantial barriers to vertical

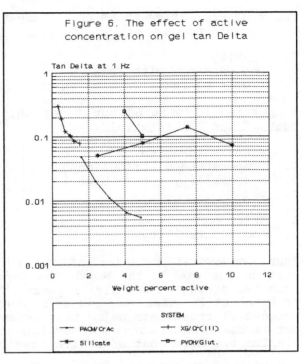

Figure 6. The effect of active concentration on gel tan Delta

migration of the water exist then the water from an aquifer or injection well may travel a large fraction of the distance to a production well in the high permeability watered out streak. The gel then diverts the water into the producing zones and the improved oil recovery may be minimal. The proposed answer to this problem is to treat the streak in depth, either from an injection well using Deep Diverting gel (DDG) compositions (30) or from a producing well using slow setting gels where the gel time is controlled by added chelants (7).

The future of water shut off and profile control treatments seems to lie in the use of the physical and chemical properties of the target reservoirs to achieve selective placement and in - depth treatment. The properties which are perceived as useful in assisting placement and those which differ between a watered out zone and the bulk of the reservoir are shown in Table H.

Table H. Properties which can be used to assist gel
          placement.
          _____

1. Gravity
2. Ion exchange with rock
3. Time (eg reaction time in chemical decomposition)

**Properties of watered out zones which differ from the
rest of the oil reservoir.**

1. Water saturation
2. Temperature distribution
3. Permeability (Pore size)
4. pH
5. Salinity
6. Brine composition
7. Rock wettability

Four commercial examples of "state of the art" gel
systems for use in the North Sea were considered here.
Each has strengths and limitations. The worldwide level
of interest in water shut off is unprecedentedly high and
successful treatments using the systems available should
act as the driving force for development of progressively
simpler and more cost effective chemical treatments.

REFERENCES.
1. US EPA Report EPA-540/1-86/035, 1984. "Health effects
    for trivalent chromium"
2. US Public Health Service Report ATSDR/TP-88-10,
    1989. "Toxicological profile for Chromium"
3. J. Doull, C.D.Klassen & M.O.Amdur, Casarett and
    Doull's Toxicology (Second Edition), MacMillan
    Publishing Co., 1980, 441 - 442.
4. Robert D. Sydansk. SPE19308 in SPE Res. Eng., August
    1990, 346 - 352. "A newly developed Chromium (III)
    Gel Technology".
5. P. Albonico, G. Burrafato, A. di Lullo &
    T.P.Lockhart. "Gelation delaying additives for
    Cr^{+3}/Polymer Gels". SPE25221 presented at the
    International Symposium on Oilfield Chemistry, New
    Orleans, LA, March 2 - 5, 1993.
6. Robert K. Prud'homme, Jonathan T. Uhl, John
    P.Poinsatte & Frederick Halverson. SPE 10948
    (10.5.83), S.P.E.J., October 1983, 804-808.
    "Rheological Monitoring of the formation of
    Polyacrylamide/Cr^{+3} Gels".
7. T.P. Lockhart, G.Burrafato & S. Bucci, Proceedings of
    the 5th European Symposium on Improved Oil Recovery.
    Budapest, 25-27 April 1989, 279-288". Cr^{3+}

Polyacrylamide Gels for Profile Modification. Crosslinking structure and Chemistry".
8. B.A.Nisbet, I.J.Bradshaw, M.Kerr & I.W.Sutherland. "Characteristics of Microbial Exopolysaccharides for EOR use". Proceedings of the Second European Symposium on Enhanced Oil Recovery, Paris, France, 8 - 10 November 1982.
9. API Recommended Procedure 63 (RP63) "Recommended Practices for Evaluation of Polymers Used in Enhanced Oil Recovery Operations", First Edition, American Petroleum Institute, Washington DC, June 1, 1990.
10. J.Liang, H.Sun & R.S.Seright, SPE 24195 presented at the SPE/DOE Eighth Symposium on Enhanced Oil Recovery, Tulsa, OK, April 22 - 24, 1992. "Reduction of Oil and Water Permeabilities using Gels".
11. Arne Stavland, Turid Ersdal, Bjorn Kvanvik, Arild Lohne, Torgeir Lund & Olav Vikane, "Evaluation of Xanthan-Cr(III) Gels for Deep Emplacement. Retention of Cr(III) in North Sea Sandstone Reservoirs". Proceedings of 7th European Symposium on Improved Oil Recovery, 27 - 29 October, 1993, Moscow, Russia.
12. J.C.Phillips, J.W.Miller, W.C.Wernan, B.E.Tate & M.H.Auerbach, Soc Pet Eng J, 25, 594 - 602, 1985. "A New High Pyruvate Xanthan for Enhanced Oil Recovery".
13. F.Lambert, M.Milas & M.Rinaudo, "Structure et Proprietes d'un Polysaccharide utilise en RAP, le Xanthane" Proceedings of the Second European Symposium on Enhanced Oil Recovery, Paris, France, 8 - 10 November 1982.
14. Matthew L. Marrocco, US Patent 4,498,540 February 12th 1985. "Gel for retarding water flow".
15. R.D.Sydansk, SPE/DOE 20,214 presented at the 7th Symposium on EOR, Tulsa, OK, April 22 - 25, 1990. "Acrylamide - Polymer/Chromium (III) - Carboxylate gels for Near Wellbore Matrix Treatments".
16. Max.F.Bechtold, Robert D.Vest & Louis Plambeck Jr., J. Am. Chem. Soc., 90(17), 4590 - 4598, August 14, 1968. "Silicic acid from tetraethyl silicate hydrolysis. Polymerisation and properties".
17. R. K. Iler, Surface Colloid Sci., 6, 1 - 100, 1973. "Colloidal Silica".
18. R.Kristensen, T.Lund, V.I.Titov & N.I.Akimov. Proc 7th European Symposium on Improved Oil Recovery, 27th - 29th October 1993, Moscow, Russia. "Laboratory Evaluation and Field Tests of a Silicate Gel System Aimed to be Used under North Sea Conditions".
19. C.L.Woods "Review of Polymers and Gels for I.O.R. applications in the North Sea", HMSO, London, 1991.
20. R.K. Prud'homme & J. T. Uhl. SPE 12640, Proceedings of the SPE/DOE 4th Symposium on Enhanced Oil

Recovery, Tulsa Okla., 15 - 18th April 1984. "Kinetics of Polymer/Metal ion gelation".

21. D. S. Jordan, D. W. Green, R. E. Terry & G. P. Willhite. SPE 10059, Presented at the 56th Annual Fall Technical Conference and Exhibition of the Society of Petroleum Engineers of AIME, San Antonio, Texas, October 5 - 7, 1981. "The effect of Temperature on Gelation Time for Polyacrylamide-Chromium (III) Systems."

22. C.B.Hurd and H.A.Letterton, J Phys Chem, 36, 604 - 615, 1932. "Studies on Silicic Acid Gels".

23. C.S.McCool, D.W.Green & G.P.Willhite, SPE/DOE 17333 presented at the SPE/DOE EOR Symposium, Tulsa, OK, April 17 - 20, 1988. "Permeability reduction mechanisms involved in In - Situ gelation of a Polyacrylamide / Chromium(VI)/Thiourea system".

24. Luc Marty, D.W.Green & G.Paul Willhite, SPE 18504 in SPE Res. Eng., May 1991, 219 - 224, "The effect of flow rate on the in - situ gelation of a Chrome/Redox/Polyacrylamide system".

25. T.Lund & A.Stavland, Proc. MINCHEM 92. The fourth Symposium on Mining Chemistry, Kiev, Ukraine, 6 - 9 October, 1992. "Influence of shear on gelation mechanism and injectivity".

26. T.S.Young, J.A.Hunt, D.W.Green & G.P.Willhite, SPE 14334 presented at the 60th Annual Tech. Conf. & Exhib. of SPE, Las Vegas, NV, September 22 - 25, 1985. "Study of equilibrium properties of Cr(III)-Polyacrylamide gels by Swelling measurement and equilibrium dialysis".

27. J.R.Gales, T.S.Young, G.P.Willhite & D.W.Green, SPE/DOE 17328 presented at the SPE/DOE EOR Symposium, Tulsa, OK, April 17 - 20, 1988. "Equilibrium Swelling and Syneresis properties of Xanthan Gum - Cr(III) gels".

28. John D. Ferry, "Viscoelastic properties of polymers", Third Edition, John Wiley & Sons Inc. New York, 1980.

29. H. Frampton, "Methods of using gel systems for Production well water shut - off treatments". Proceedings of PSPW Technical Seminar, Csarna, Poland, 15th June 1993.

30. A.J.P.Fletcher, S.Flew, I.N.Forsdyke, J.C.Morgan, C.Rogers & D.Suttles. J. Petroleum Sci. & Eng., 7, 33 - 43, 1992. "Deep diverting gels for very cost-effective waterflood control.

New Water Clarifiers for Treating Produced Water on North Sea Production Platforms

D. K. Durham

BAKER PERFORMANCE CHEMICALS INC, HOUSTON, USA

INTRODUCTION

Regulations applying to produced water discharge in the North Sea have become increasingly stringent and it is difficult for many platforms to remain in compliance with respect to oil content in the overboard produced water.

With increased water production most platforms are equipped with water treating systems that have difficulty treating the current volume of produced water down to 50 ppm and no sheen, using conventional water clarifiers.

Because of the increased water production and stricter enforcement by the regulatory agencies, producers in the North Sea were facing the possibility of being out of compliance unless new water treating equipment could be installed, which for many platforms would be very expensive due to space restrictions and construction costs.

Because of this situation a product development program was initiated to develop new water clarifiers that would allow the existing offshore water treating systems in the North Sea to treat the overboard water to meet the governmental specifications pertaining to the overboard discharge of produced water.

The initial results were the first generation dithiocarbamate (DTC)-type water clarifiers $\{SC(:S)N(R')RN(R')C(:S)S^-\}$ that were effective in water clarification but produced an unmanageable "floc" that in most production systems created severe operational problems. The term "floc" refers to the substance formed when water clarifiers remove oil and solids from produced water. This floc contains primarily oil, solids, and water clarifier, and is normally in the form of an amorphous, water insoluble paste that has limited oil solubility depending to a large extent on the type of water clarifier used to form the floc.

Because of the floc-related problems, second and third generation dithiocarbamate type water clarifiers were developed which provided floc characteristics that were manageable within the existing systems in contrast to the unmanageable floc produced by the first generation dithiocarbamate-type water clarifiers.

PRODUCT APPLICATION/SYSTEMS REVIEW

Due to the 40 ppm limit there was a need for improved water clarifiers or improved water treating systems, especially for the platforms for which water production had increased to the point where the existing equipment was inadequate for treating the water.

In addition, some platforms exhibited a visible slick below the 40 ppm limit. In many cases DTC-type clarifiers were the only type of products that were capable of consistently treating the waters to meet the governmental criteria to the existing water treating systems.

The systems and platforms vary widely with respect to retention time and equipment type but most have either two or three types of separation vessels where water can be treated. The three types of treating equipment include primary separation equipment, gravity settling equipment, and flotation equipment.

Previously, most platforms were treated by injection of water clarifier prior to either gravity settling or flotation. With stricter adherence to the 40 ppm limit treatment in mixed production, gravity settling, and flotation either individually or combined, was initiated which in some cases allowed operators to meet the 40 ppm/no sheen criteria. In most cases, however, the amount of chemical used and floc-related mechanical problems increased dramatically. These problems were especially noted where skimmers (gravity settling vessels) were used, due to their inability to operate automatically at levels where floc can be skimmed continuously and recycled through the system. This results in the buildup of large quantities of floc in the skimmers, which is recycled through the system over a short time frame, causing floc to build in treating vessels, and in many cases cause wet oil.

In most mixed production and flotation water treatment, the floc is recycled into mixed production on a continuous basis allowing the floc to dissolve or disperse into the oil phase. The use of hydrocyclone and centrifuge-type gravity settling equipment could resolve the floc recycle problem associated with other gravity settling equipment.

Floc-related problems can occur with both conventional and DTC-based clarifiers and in both cases can be reduced by insuring that floc is continuously recycled into the systems allowing maximum retention time and dissolution into the oil phase. Conventional products are not fast or efficient enough in reducing the oil count in low residence time systems to adhere to the 40 ppm limit.

Conventional products include metal salts, metal salt blends, anionic and cationic solution polymers, polyamines, latex polymers, and emulsion polymers. Because the existing conventional products could not resolve the problem, product development efforts were increased to identify more effective water clarifiers that would allow existing systems to effectively treat the produced water.

DITHIOCARBAMATE WATER CLARIFIERS

Water clarification technology was searched and a synthesis program was conducted to investigate new and existing water clarifier chemistry that might achieve the desired performance.

After testing a wide range of water clarifier chemistries, certain type of organic dithiocarbamates were identified as effective first generation DTC-type water clarifiers.

Dithiocarbamate applications noted in the literature include ore flotation collectors in the mining industry and chelating agents for removal of various metals from plating solutions prior to discharge or reuse.

Laboratory and field test data show that DTC-type water clarifiers function by reacting quickly with iron and other dissolved metals to form a floc that quickly coalesces and removes oil from produced water. The speed of the reaction of the DTC with iron to initiate floc formation allows the DTC water clarifiers to treat water much faster than conventional products.

In addition, DTC-type clarifiers were found to be capable of achieving a much lower insoluble oil count compared to conventional products. This becomes an important advantage where a sheen is produced at low parts per million insoluble oil.

Another important advantage found with DTC water clarifiers is that they are in many cases equally effective in either mixed production, gravity settling, or flotation on the same platform. Most conventional products are normally effective in only one of the three types of application modes.

FIRST GENERATION DTC WATER CLARIFIERS

After the first generation DTC water clarifiers had been in use for a short time it was apparent that the floc generated by these products was, in most facilities, unmanageable and caused severe mechanical problems.

This floc caused severe problems due to its adherence on metal surfaces and floc build-up at interfaces in separation and water treating equipment.

The primary floc-related problems included mechanical fouling of float cages, level controllers, plugging of float cell skimming troughs, and interface build-up in treating vessels.

The floc problems generated by these first generation DTC clarifiers was caused by the lack of solubility of the oily floc in the produced oil. Because of these problems further synthesis and testing was conducted to develop DTC water clarifiers that would produce a manageable floc.

SECOND GENERATION DTC WATER CLARIFIERS

Synthesis and testing conducted with focus on floc characteristics resulted in the identification of second generation DTC water clarifiers which, when used in the treating systems, produced a floc that was manageable in most of the systems but did not totally eliminate long term floc related problems in some systems. This was a significant improvement and allowed operators to meet oil content specifications and operate for the most part without unexpected floc-related failures. Attachment 1 illustrates the comparison of first and second generation DTC type clarifiers with respect to floc-related mechanical problems.

The second generation products were also found to be highly effective in mixed production, gravity settling and flotation application modes. However, mixed production and flotation application have proven to be the most effective and manageable application modes for both DTC and conventional water clarifiers. The reason for this is that most gravity settling applications require the use of higher concentrations of clarifier which generate more floc and accelerate problems related to floc accumulation and floc-related failures.

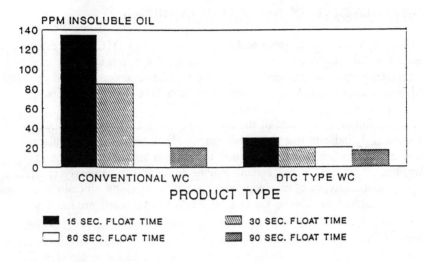

PLATFORM B TEST DATA
BENCH MODEL FLOAT CELL

60 PPM TREATING RATE

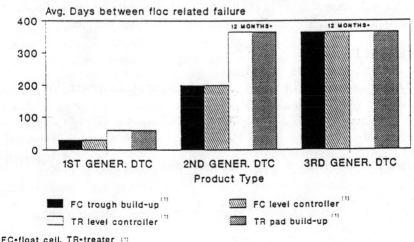

PLATFORM B PLANT TEST DATA
Comparison of DTC Floc Manageability

Attachment 1

In most conventional gravity settling equipment (skimmers, CPI, tanks) floc problems are aggravated by the fact that the oily floc is not continuously recycled back into the system but is sporadically batched into the system at high concentrations causing system upsets.

THIRD GENERATION DTC WATER CLARIFIERS

Development work in the area of floc improvement of DTC water clarifiers has resulted in third generation DTC-type water clarifiers that are approaching the floc manageability of conventional products while retaining most of the speed and effectiveness of first and second generation DTC based clarifiers.

Preliminary test data of the third generation DTC clarifiers, shown in attachment 1, indicate that they have almost eliminated floc accumulation problems in treating vessels compared to first and second generation DTC clarifiers. With the discharge requirements becoming increasingly restrictive, the use of these types of products may become a necessity since in many cases they are required to achieve the low oil levels in overboard produced water.

SUMMARY

First generation dithiocarbamate-type water clarifiers produce a difficult to manage floc that creates unacceptable mechanical problems for operators, although the desired water clarification ability is achieved.

Second and third generation dithiocarbamate water clarifiers have, in most cases, replaced the first generation products because they are equally effective with respect to water clarification and eliminate the severe floc-related problems associated with the first generation DTC water clarifiers.

REFERENCES

van Oss J.F., Chemical Technology; an Encyclopeodic Treatment, Vol. 1, Barnes and Nobel, 1968.

Gaudin A.M., "Flotation", McGraw-Hill, New York, 1957.

Adamek E.G. and Hudson G.B., Dithiocarbamate Ore Collectors, CAN 771,181, 1967.

Siggia S., Quantitative Organic Analysis via functional groups, Wiley, New York, 1967.

Subsurface Disposal of a Wide Variety of Mutually Incompatible Gas-Field Waters

G. Fowler

NEDERLANDSE AARDOLIE MAATSCHAPPIJ BV, BUSINESS UNIT GASLAND, SCHOONEBEEK, NETHERLANDS

OVERVIEW

The Netherlandse Aardolie Maatschappij bv is divided into Business Units, the main ones being:

B.U. Groningen (BUG)   Responsibility for the Groningen Onshore Gas field.

B.U. Gas Land (BUGL)   Responsibility for the numerous "smaller" ONSHORE gas fields within the Netherlands.

B.U. Oil (BUO)   Responsibility for the Oil field in Schoonebeek and the Rotterdam geographical areas.

B.U. Offshore (BUS)   Responsibility of the Offshore Dutch North Sea Sector Gas and Oil fields.

B.U. Services   Responsible for the provision of services in drilling, supply, waste disposal, environment, chemistry etc.

B.U. Gas Land operates 41 "smaller" onshore gas fields in which there are approximately 200 locations of activity where, potentially, water may originate for disposal. The activities on individual locations can range from drilling & workover, gas & condensate production, pipeline cleaning, suspended locations and water disposal operations.

The type of water which requires disposal varies greatly from run-off rainwater to super-saturated connate waters. The volume per field or installation also varies greatly from a minuscule 30 m³ per year to a significant 20,000 m³/year with one location producing potentially 150,000 m³ per year. With the exception of the latter case, the volumes requiring disposal are too small per installation to make pipeline transfer a cost-effective viable transportation system. Although geographically widespread, the availability of suitable disposal zones precludes the development of independent water disposal installations.

Prior to July 1992 BUGL had two injection locations of its own available for disposal of the waters.[1]
One location, was dedicated to waters from the "sour" gas fields and injected the water into the highly permeable sandstone, itself a "sour" oil producing zone in the Schoonebeek Oil field.
A second location, was reserved for the "sweet" waters originating from the South East Drenthe Fields i.e. the geographical area bounded by the triangle from Roswinkel to Hardenberg to Wanneperveen.

Waters from North West Drenthe were processed by the B.U. Oil via a section of their treatment system. Waters from the Friesland area were disposed of by B.U. Groningen. Additional disposal was achieved near Amsterdam and in Rotterdam Area by the BU Offshore and BU Oil respectively.

In early 1992 several events happened simultaneously and effectively restricted the disposal of water via the historical outlets to such a degree that a fresh approach was needed for the continued long-term disposal of BUGL water. A phased implementation was devised because of the short lead time for equipment procurement, NAM required to have some form a water treatment available quickly and the interaction of the various waters was not fully evaluated.

EAST NETHERLANDS WATER DISPOSAL (ENWD) PHASES

The phases defined in 1992 were:

pre-Phase 1   Define a suitable location for the installation and the minimum process[2,3] required for suitable water treatment.

Phase 1A   De-bottleneck the selected installation and construct a pipeline from it to the BUO Injection Installation

Phase 1B   Construct the Pilot (minimum) treatment installation on the selected installation.

Phase 2   Construct a Mechanical Vapour Compression unit at the BUO Injection Installation.

Phase 3   Construction of the treatment system utilising the equipment and experience gained in Phase 1B operation[4].
Construct a pipeline and injection facilities on an alternative BUGL depleted gas reservoir.

During the operation of the Phase 1B, several problems were encountered which required the allocation of an ENWD Phase 4.

Phase 4   Inclusion of a sludge compaction system
Interstage treatment of fluids originating from the Gas Adsorption System on the selected installation

During the Phase 1 and 3 operational and design periods, several factors were identified which had an impact upon the scope of design of the East Netherlands Water Disposal Scheme :

The condensate based corrosion inhibitor, which has been used for the last 10 years, was found to promote the formation of an extremely stable condensate-in-water emulsion in the produced water originating from the Gas Adsorption System.

The incompatibility of the waters produced a variety of precipitated solids in the form of metallic sulphides, metallic & alkaline earth sulphates and carbonates.

The formation waters were highly unstable requiring extensive site analysis and ionic species fixing to obtain dependable and reproducible water analyses.[5,6]

Sulphate Reducing Bacteria were suspected to have become a problem based upon sessile bacterial studies during the Phase 1B period. This was actually found to be the situation and a comprehensive biocidal treatment regime was defined.

Low Specific Activity (LSA) Scales, also termed Naturally Occurring Radioactive Materials (NORM), were known to originate from several fields and BUG made a decision to preclude these waters from their disposal systems.

After an evaluation of suitable disposal venues[2], the decision was made to locate the ENWD Treatment Facility at Schoonebeek [3]. The reasons for selecting this location were quite diverse and were as follows:

> There was an existing Sour Gas Treatment Facility with a relatively large 600 m^3 produced water storage tank.

> Relatively close proximity to the existing water injection ring main system where several dedicated wells could be removed from the ring main system.

> High volume water producing gas wells were processed in the Gas Absorption Installation on the selected installation.

> The selected installation was already a designated LSA susceptible installation.

> Water delivery trucks could achieve easy access with the minimum environmental impact.

> The selected installation was geographically well situated for utilising depleted gas reservoirs in the future.

> Building and Operating Permits would be easier to obtain for an existing location.

> Space was available to extend the existing installation plot plan.

Table 1 gives a summary of the water chemistry ofsome of the various types of water which have been handled during the first 18 months of operation.

CONTAMINATED RAINWATER

The run-off rainwater requires subsurface disposal due to the extremely strict environmental surface discharge in force within the Netherlands and the undertaking of NAM to operate with the minimum negative environmental impact. The discharge (overboard) water to the surface drainage canals and ditches may not contain more than 200 µg/l of total hydrocarbons, 2 µg/l BTEX (Benzene, Toluene, EthylBenzene, Xylene) and 0.05 µg/l Mercury[7] to name but a few components. In addition, NAM does not discharge surface waters into areas of particular environmental sensitivity or domestic water abstraction. It was quickly ascertained during the first six months that more than 20% of the water truckedfor disposal was from the rain water pits. This represented only a fraction of the uncontaminated rain water which was routinely disposed of via the surface canals and ditches.

It can be seen that this represents an enormous cost implication in water management to NAM. For example for every cubic metre of rainwater injected into the disposal zone there is a deferment of 1 cubic metre of produced formation water which is associated with a several orders of magnitude of saleable gas.

Another implication of injecting surface rain water is that the Netherlands Law and Mining Regulations states that "reservoir foreign water and material "may not be disposed of by re-injection. NAM has a dispensation[1] to inject a pre-defined volume of rainwater but it is possible that this dispensation may be reduced or removed in the future.

CONNATE & PRODUCED WATERS

The philosophy of the ENWD scheme has been to develop a process which will be as flexible as is technically feasible in order to accept the widest variety of mutually incompatible waters.

In Table 1 it can be seen that the waters originating from the sour gas fields contain hydrogen sulphide and mercaptan concentrations up to approximately 100 mg/l. Waters originating from several fields contain up to 620 mg/l dissolved *Iron*. When these waters co-mingle precipitation of solid *iron sulphide* occurs. In many water deliveries to the treatment system there was 1-5% by volume of free hydrocarbon (condensate). The run-off water from the rainwater pits contains 6-8 mg/l of dissolved oxygen and this must be removed for corrosion control and to prevent the oxidation of soluble cations to insoluble hydroxides.

TABLE 1   SUMMARY ANALYSES OF SOME OF THE WATER TYPES

	#1	#2	#3	#4	#5	#6	#7	#8	#9
				Water Chemistry Type (ionic concentrations in mg/l)					
Water Type	Connate	Connate	Produced/ Equilibrium Mixture	Connate	Connate/ Equilibrium	Connate/ Diluent	Produced/ Equilibrium Mixture	Produced Mixture	Contaminated Rainwater
Pressure pH	6.8	5.8	6.0	4.1	5.2	4.8	5.1	5.0	6.8
Temperature	20.0	15.0	18.0	20.0	34.0	50.0	18.0	46.0	10.0
Total Sulphides	< 0.5	0.0	< 1	0.0	13.0	0.2	0.5	102.0	0.0
Dissolved CO2	150.0	427.0	29.0	300.0	600.0	154.0	243.0	686.0	1.0
Dissovled O2	< 5 ppb	< 5 ppb	< 5 ppb	< 5 ppb	< 5 ppb	< 5 ppb	< 5 ppb	< 5 ppb	6.0
Chloride	95000.0	58000.0	17.0	180000.0	8000.0	157000.0	4300.0	117000.0	100.0
Sulphate	2800.0	1120.0	64.0	120.0	18.0	50.0	10.0	100.0	50.0
Bicarbonate	280.0	187.0	17.0	27.0	63.0	11.2	10.0	40.0	16.0
C1-C5 Org Acids	< 5	44.0	309.0	80.0	162.0	8.0	23.0	11.0	3.0
Sodium	58000.0	28000.0	32.0	84000.0	1200.0	72000.0	1320.0	40500.0	81.0
Calcium	2000.0	15000.0	49.0	26000.0	1570.0	21200.0	144.0	24100.0	23.0
Strontium	77.0	121.0	0.2	1100.0	37.0	600.0	3.3	1220.0	< 1
Barium	< 1	4.0	0.1	24.0	2.3	60.0	0.2	34.0	< 1
Total Iron	27.0	220.0	21.0	370.0	620.0	28.0	57.0	34.0	3.0
Mercury	< 0.01	< 0.002	0.0	0.0		0.0	0.0	0.0	< 0.001
Lead	< 1	2.0	< 0.002	80.0	0.1	16.0	0.0	4.8	< 0.001
Zinc		1.0	0.4	13.0	1.7	2.5	0.4	38.0	0.0
Methanol	0.0	0.0	16100.0	0.0	0.0	0.0	2280.0	0.0	20.0
Glycols	0.0	0.0	3500.0	0.0	0.0	0.0	1060.0	0.0	40.0
Scale Observed	CaCO3 FeCO3	CaSO4 FeCO3		BaSO4 CaCO3	FeS CaCO3	FeS RaSO4 PbSO4 BaSO4	CaCO3	FeS ZnS PbSO4 RaSO4 PbS CaCO3	
LSA Status	Normal	Normal	Normal	LSA	Normal	LSA	Normal	LSA	Normal

Ammonium Bisulphite Oxygen Scavenger is injected into the inlet of Tank T-12 with every road tanker water delivery, not only to reduce the corrosion rate, but mainly to prevent the. deposition of *ferric oxides* and *hydroxides* as well as *elemental sulphur.*

Some of the connate waters have high concentrations (up to ±3,000 mg/l) of *Sulphate* whereas other waters are rich in *Strontium* and *Barium* (up to 1500 mg/l Sr & 60 mg/l Ba). Almost all the connate waters from every field have high concentrations of *Calcium.* Many of the produced waters which require disposal are connate waters diluted by condensed water vapour giving a much lower ionic composition.

Precipitation of the *sulphates* of *calcium, barium* and *strontium* are therefore highly probable and in order to reduce the possibility of scale deposition causing line blockage, equipment failure and reservoir impairment, a phosphonate based scale inhibitor was injected at the inlet to Tank T-12 and the discharge from low pressure separator.

The philosophy of injecting scale inhibitor is not the total prevention of scale precipitation, as this would probably be difficult to achieve, but the reduction of post - filtration precipitation.

Some of zones in the gas producing reservoirs contains the natural decay series of *Uranium-238* (^{238}U) and *Thorium-232* (^{232}Th) mineral and the radiological decay over the ages has produced the radioactive isotopes of ^{210}Pb, ^{226}Ra, ^{228}Ra and ^{222}Rn amongst others[7,8]. Although the radioactivity concentration may be relatively high, the mass of these radionucleides can be measured in terms of nanogrammes.

A scale with an observed activity of 160 Bq/g ^{226}Ra would correspond to a minuscule 0.009 µg/g of actual radioactive $^{226}RaSO_4$ *(Radium Sulphate)* scale mass.
The question must be asked of the scale inhibitor vendors whether it would be practical to prevent the formation of such microscopic amounts of scale by using the current scale inhibitor chemistry or whether a new range of products must be developed to maintain a radioactive free system.

In many of the connate waters there are significant concentrations of *Lead,* present as the radioactive (^{210}Pb, ^{211}Pb, ^{212}Pb, ^{214}Pb) and non-radioactive isotopes^{204}Pb, ^{206}Pb, ^{207}Pb, *Zinc* and *Mercury.* Given the fact that the conditions of ionic incompatibility exist in the ENWD water treatment process for the precipitation of these elements as sulphides, sulphates and carbonates, then it came as no surprise that these compounds were identified in the scales and sludges throughout the BUGL Gas and Water Systems.

Acceptance of the concept that precipitation of toxic, radioactive and reservoir plugging solids would occur and could not be prevented by chemical inhibition had a fundamental impact up on the process design philosophy as discussed later in this paper.

The water chemistries and incompabilty interactions which are discussed above have many similarities to those reported by other operators in the North Sea and elsewhere.[8,9,10,11]

MISCELLANEOUS FLUIDS

This category was created to include the small volume , when compared to connate and rain waters, aqueous streams originating from diverse operations such as wirelines, workovers, welltests, process vessel cleaning, pigging operations and emulsion generation.

As can be appreciated, these sources can give conditions where the treatment process is exposed to transient conditions of high suspended solids, water-in-condensate emulsions, high or low pH values etc. Because of the limited facilities which were available for Phase 1B in connection with sludge removal and solids generation, the waters from these diverse activities were precluded. These difficult fluids were, however, accepted after presettement at another installation which had 2 eighty cubic metre tanks available as well as demulsifier injection and recirculation pumps. In Phase 3 a dedicated handling and treatment system was designed which incorporated a heating, demulsifier chemical injection and recirculation loop for demulsification, as well as a conical bottom tank for sludge removal.

SLUDGE TREATMENT

In Phase 1B, the decision was taken not to accept fluids containing high concentrations of suspended solids or high levels of H_2S. This was to limit the sludge generation and removal aspect. In Phase 3 , however, there will be a significant amount of solid material which will need to be removed on a regular basis.

The safe handling and disposal of the sludge from the ENWD Treatment Plant has caused a significant problem during the design of Phase 3.

A concept was developed where the sludge containing 15 - 20% by volume solids could be removed from the conical bottomed tanks in Phase 3 and pumped to a V-Sep™ Unit where the sludge could be thickened to 70-80% vol. of solids. V-Sep, which stands for Vibratory Shear Enhanced Processing, was developed by New Logic International[13] of California and is designed to overcome the fouling problems associated with membrane system technology. The thickened sludge will then discharged into sealed drums for disposal using a company which specialises in the handling, transportation and disposal of toxic or radioactive waste material. At no time will the personnel be exposed to the sludge itself. Radioactivity will be continuously monitored in the discharge stream. In Phase 1B, sludge was obtained for test purposes and used to check the viability of the concept. The sludge was passed through a membrane head containing various types of membrane material. It was found that one particular membrane allowed the concentration to the desired consistency and was not blocked by the particulates in the sludge itself. In addition the membrane reduced the salt content of the water but this is an unnecessary aspect for this particular process.

Further longer duration tests are planned to evaluate the long-term fouling characteristics of the membranes.

As a significant concentration of *iron sulphide* will be in the sludge the risk of auto-ignition from pyrophoric iron had to be considered. This is the reason that the thickened sludge still contains 20% by volume of water. Analysis of the sludges originating in Phase 1B and from various segments of BUGLs' operations indicates that *metallic mercury, organo-mercury, inorganic mercury salts, lead, cadmium, zinc sulphides, strontium, barium sulphate scales, radium* and *lead sulphates* and *sulphides, hydrocarbon condensate* and *aromatics* (benzene, toluene etc.) could also be present in the separated solids.

This presents a significant health and safety hazard[12] which is why no personnel will be physically exposed to the sludge and the disposal and transportation will be undertaken by specialised and certified contractors.

PROCESS DESIGN AND PERFORMANCE

As well as being a self supporting treatment system, Phase 1B was also a pilot plant used to define what treatment stages would be required for the future Phases. In the start-up stage of Phase 1B, an induced gas flotation unit and a mixed nutshell media filter were included. The units were obtained on a rental basis and could be returned if unnecessary.

Figure 1 shows the flow scheme of the Phase 1B process and Figure 2 is the process defined for Phase 3. The necessity for the induced gas flotation unit was verified for the Phase 3 process to reduce the impact of periodic high concentrations of condensate and the continuous suspended sub-micronic deposits of FeS on the nutshell media filter. It was also decided that in the Phase 3 the inclusion of the mixed nutshell media[14] filter would be required to remove residual condensate and solids. Tank T-12 was an existing tank on the selected location.
It was decided to use this tank in Phase 1B as the reception tank for water trucked in from the various fields.

Tank T-12 was used to allow a time of 24 - 36 hrs for the incompatible waters to react and precipitate any scales or heavy (sands) suspended material. Because of the need to keep Phase 1B as simple as possible and the limited time it would be in use, it was decided to recycle the filter backwash fluids back to T-12, and restrict the types of incoming water to limit the amount of H_2S and hence the generation of FeS. It was calculated that there would be enough sludge capacity in the flat bottomed tank, T-12, to allow operation for 20 months.
As the H_2S containing waters were precluded and it was found during commissioning that the condensate settlement in Tank T-12 was effective, it was decided to decommission the Flotation Unit and operate Phase 1B with only the Media Filter.

Monitoring of the Total Hydrocarbons and Total Suspended Solids content on a daily basis gave a measure of the system performance as a function of time and water types. After 17 months continuous usage, it was found that the filter back-wash frequency had to be reduced from 4 hours to 3 hours then to 2 hours.
This was the result of the build up of the recycled sludge plus an increased solid generation due to sulphate reducing bacteria producing excessive amounts of H_2S. The implementation of Phase 3 has been deferred from October 1994 until mid 1995 and therefore it became necessary to clean Tank T-12. to allow the continued operation of Phase 1B until mid 1995.

During the design segments of Phase 1B and Phase 3, the decision was taken to by-pass the water treatment system with the fluids originating from the Gas Adsorption System. The basis of the decision was that this particular stream constituted $1/2$ to $2/3$ of the total volume envisaged for the ENWD, which in volume terms, is equivalent to 300 to 1000 m^3/ day of water. The water analysis indicated that the water was of a good enough quality for injection with only gravity settlement of the condensate after the co-mingling of the treated trucked water and untreated Gas field water.

During the operation of Phase 1B, it was found that the residence time in the treated water Tank T-15 was insufficient to remove the entrained condensate. During 1993 it was necessary to increase the corrosion inhibitor injection rates into the some gas wells. The result was that the self imposed treatment specification of 40 mg/l suspended hydrocarbon was constantly exceeded by an order of magnitude. Additional tank volume was available just prior to the injection pumps which allowed further condensate removal. However, the

FIGURE 1. PROCESS FLOW SCHEME FOR ENWD PHASE 1B & TESTING

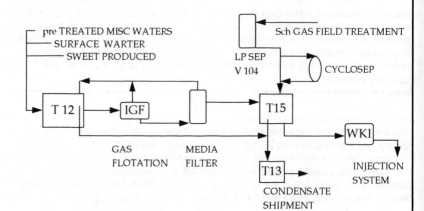

pre TREATED MISC WATERS
SURFACE WARTER
SWEET PRODUCED

Sch GAS FIELD TREATMENT

LP SEP
V 104

CYCLOSEP

T 12   IGF   T15

GAS   MEDIA   WK1
FLOTATION   FILTER

T13   INJECTION
                             SYSTEM
CONDENSATE
SHIPMENT

FIGURE 2.   PROCESS FLOW SCHEME FOR ENWD PHASE 3

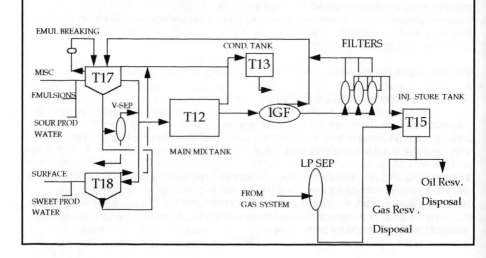

EMUL. BREAKING

COND. TANK   FILTERS

MISC   T17   T13

EMULSIONS   V-SEP   INJ. STORE TANK

SOUR PROD   T12   IGF   T15
WATER
                 MAIN MIX TANK

SURFACE   T18   LP SEP
                                                         Oil Resv.
SWEET PROD   FROM   Disposal
WATER   GAS SYSTEM   Gas Resv .

                                                Disposal

hydrocarbon in the injected water was still well above the water quality specification. During the first 20 months of injection into the highly permeable and porous Sandstone Reservoir, there was no injectivity impairment observed. However, this situation could not be considered as ideal when future disposal requirements were considered. Replacement disposal reservoirs were to be needed in the future and Phase 3 ENWD included the laying of a pipeline and installation of injection pumps at a second location for disposal into a fractured carbonate reservoir.

The water quality requirements for the carbonate reservoirs are much higher than that acceptable to the sandstone. One of the options available was install a condensate removal unit between the Low Pressure Separator, V-104 and Tank T-15. It was decided to evaluate a unit which comprised of a combination induced gas flotation and hydrocyclone, The unit theoretically required a much lower inlet pressure than conventional hydrocylones[15]. A Cyclosep™ from Monosep Corp. was procured on a rental basis for a three month trial.

Mechanical operation was perfect but the unfortunately the condensate removal efficiency was only about 50%. Investigation into the poor performance revealed that the fluid stream coming from the V-104 was 99.5% vol. water and 0.5% vol. Condensate + Corrosion Inhibitor mixture. This would not have posed a problem to the Cyclosep had there been two discrete phases but unfortunately the fluid stream consisted of 100% vol. of condensate-in-water emulsion which was generated by high shear conditions across the well chokes and vessel control valves[16].

Bottle testing with over 150 products from different vendors has failed to find a suitable demulsifier, polyelectrolyte or clarifier. The emulsion was extremely stable, with separation to a clear water phase occurring after 12 hours of passive conditions. The continuous high values of hydrocarbons in the injection water stream was due to the entrainment of the unresolved emulsion.

In an attempt to solve the emulsion problem at the cause rather than treating the effect, a study was started to find a corrosion inhibition system which will not require injection of condensate as a diluent and will not generate or stabilise emulsions. The corrosion inhibitor study was to be undertaken in any event as part of a drive to investigate the corrosion mechanisms and inhibition in the sour gas wells. Additional requirements include finding an inhibitor which has a lower environmental toxicity and can be incorporated into a corrosion/scale inhibitor combination product.

FUTURE DEVELOPMENTS.

METHANOL CONTAMINATED PRODUCED WATERS

There are limits on the concentration of the production chemical residuals which can be present in the disposed water. From two locations in the Northern Netherlands there is a very low salinity water containing 1-15% *Methanol*. The water is number 3 in Table 1 and is predominantly condensed equilibrium water. The methanol is present because of its use in some wells as a gas-hydrate inhibitor This water exceeds the government legislation and the operating permits for the ENWD in respect of the methanol concentration. Disposal is currently via a third party specialist effluent treatment company who distil off the methanol prior to water disposal in their own treatment system. The regenerated methanol is returned to NAM for re-use.

A significant cost saving can be achieved and NAM will increase its water management integration if this water/methanol mixture could be treated "in-house" prior to disposal. There is an investigation being considered where the methanol is removed from the water by the use of a vibratory enhanced membrane separation technique (V-Sep) and returned to bulk storage. The water could then be transferred to ENWD by truck for sub-surface disposal. Apart from the external reprocessing costs which could be saved, another benefit of this technology would be to comply with requirements by the Netherlands Government to utilise the "Best Available Technology" to maintain an environmentally acceptable operation.

BIODEGRADATION OF CONTAMINATED RAINWATER

As a parallel investigation, which has an impact on the ENWD phases, there is a project being undertaken to develop a low capital cost, low operating cost biodegradation system which will treat rain water to meet the environmental discharge legislation.

The system will need to be retrofitted to every water pit (over 200 in BUGL).

This will free a significant amount of reservoir voidage for produced water disposal. Additional free voidage is critical to the ENWD for several operational and reservoir reasons.

A field test unit has already been used to evaluate whether biodegradation to the very low limits required for disposal is viable.

The concept has proved satisfactory and the second stage of the study was to commence in April 1994 to install a modified cell immobilised bacterial strain in the test unit. An objective is to achieve biodegradation within a single pass of the unit. Chemical injection of a special chelant/flocculant will also be incorporated to remove the trace concentrations of the heavy metals present in the water.

SOIL REMEDIATION

NAM abandons its land based installations when a particular reservoir section is depleted. Some of these installations have been in constant operation since 1947 and consequently there has been significant contamination of the top soil strata by hydrocarbons and salt. NAM undertakes to clean up the soil before returning them to the original owners. The clean-up of each installation can take several years and produce hundreds of thousands of cubic metres of mildly contaminated ground water.

One option for disposal of the ground water would be re-injection via the ENWD system. This is really not an economic or environmentally viable option because of the large volumes which are envisaged. An alternative solution would be the treatment of the water at the remediation site and disposal to the surface canal systems.
Possible technologies which could be used are biodegradation of the organics, chelation/flocculation of the heavy metals and membrane technology for salt removal[11,17,].

CONCLUSIONS

The sub-surface disposal of chemically incompatible waters has been shown to be a viable solution to the disposal of a wide range of oil and gas field waters.

The inter relationship of different segments of gas field operations, environmental, health, safety and legislation can have a high impact on the overall water management philosophy.

The development of an inflexible concept at the outset of the disposal system design would have a detrimental effect on the overall concept of water disposal. particularly in situations of diverse water sources.

ACKNOWLEDGEMENT.

The author wishes to thank the Nederlandse Aardolie Maatschappij for permission to present this paper and in particular to the Operations Support and Facilities Engineering Groups for the assistance in developing the various phases of the East Netherlands Water Disposal Project.

REFERENCES

1.   Marquenie, J.M., Kamminga,G., Koop H., Elferink T.O. "Onshore Water Disposal in the Netherlands: Environmental & Legal Developments" SPE Paper Nr 23320 Presented at First Int'l Conference on Health, Safety & Environment, Den Haag, Netherlands, 10 - 14 November 1991.

2.   Robinson, D.R., Oil Plus Ltd., Specially Commissioned Evaluation, "Report on the options for Treatment of Waste Waters in a Central Facility" Report Nr. ES 4270A, Sept. 1991.

3.   Internal NAM Report Nr 21151, "Water Disposal Investigation - Phase 1 Implications" March 1992.

4.   Internal NAM Report Nr 22832, "Commissioning Study of Phase 1B of the Water Treatment Installation & the Implications on the Phase 3 Design" Dec. 1992.

5.   Fowler, G., Gunn M., "Accurate Chemical Analysis - The basis for dependable Laboratory Studies & Process Design" Presented at UK Corrosion '90 29th October 1990.

6.   Oil Plus Ltd, Newbury England. "Site Water Analysis Procedures - On Site Fixing and Analysis of Bicarbonate, Carbonate and Carbon Dioxide Species".

7.   Minister van Verkeer and Waterstaat, "Derde Nota Waterhuishouding -. Water voor nu en later" ISBN Nr 90 12 06 353 1

8.   Stephenson, M.T. "Components of Produced Water: A Compilation of Results from Several Industry Studies" SPE Paper Nr 23313 Presented at First Int'l Conference on Health, Safety & Environment,Den Haag, Netherlands, 10 - 14 November 1991.

9.   Bassignani, A. Di Luise, G. & Fenzi, A. "Radioactive Scales in Oil & Gas Centres SPE Paper Nr 23380 Presented at First Int'l Conference on Health, Safety & Environment, Den Haag, Netherlands, 10 - 14 November 1991.

10.   Simms, K. "Recent Studies in Produced Water Management in Canada"
        Presented at the Water Management Offshore Conference, Aberdeen, Scotland.
        6-7 October 1993.

11.   Hansen, B. "Review of Potential Technologies for the Removal of Dissolved
        Components from Produced Water" Presented at the Water Management Offshore
        Conference, Aberdeen, Scotland. 6-7 October 1993.

12.   Grice, K.J. "Naturally occurring Radioactive Materials (NORM) in the Oil & Gas
        Industry: A New Management Challenge" SPE Paper Nr 23384 Presented at First I
        Int'l Conference on Health, Safety & Environment, Den Haag, Netherlands,
        10 - 14 November 1991.

13.   Culkin, B "Vibratory Shear Enhanced Processing: An Answer to Membrane
        Fouling" Chem. Proc. Jan 1991

14.   Sabey J. B., Pawar, S., "Developments in Deep-bed Filtration for Produced Water
        Re-injection & Disposal" Presented at Water Management Offshore Conference,
        Aberdeen, Scotland, 6-7 Oct. 1993.

15.   Skilbeck, F. et al "Use of Low Shear Pumps and Hydrocyclones for Improved
        Performance in the Clean-up of Low Pressure Water" J. SPE Prod. Eng.
        August 1992 p295 - 299.

16.   Gramme, P.E. "Treatment of Produced Water at Gas/Condensate Fields"
        Presented at Water Management Offshore Conference, Aberdeen, Scotland,
        6-7 Oct. 1993.

17.   Veltkamp, A.G., Mathijssen, J.J.M. "Cleanup of Contaminated Soil & Ground
        water: A Location Specific Cleanup Operation at the Sappemeer Gas Production
        Site" SPE Paper Nr 23377 Presented at First Int'l Conference on Health, Safety
        & Environment, Den Haag, Netherlands, 10 - 14 November 1991.

Subject Index

H

Haltenbanken, 125
He-Hua Zhou, 281
hemicellulose, 63
Henaway, P., 163
hexacyanoferrate, 208
Hodder, M.H., 28
Hoffman, G.G., 189
hole, collapse of, 13
Holliman, P.J., 234
Hourston, K.E., 126
Hoyle, R., 234
Hughes, T.L., 99
hydrogen sulphide, 179, 183, 190,
    202, 208, 214, 234
hydrostatic pressure, 15
hydroxyethyl cellulose (HEC), 88

I

improved oil recovery (IOR), 277
injection well temperature, 190
insurance chemicals, 277
ion exchange processes, 266
iron,
    chelated, 208
    regenerable oxidant, 207

J

Jackson, G.E., 164
Jefferies, M., 55
Jiang, P., 126
Johnson, R.W., 275
Jones, T.G.J., 99
Jordan, M.M., 126

L

Lawless, T.A., 38, 251
Laycock, P.J., 179
lignite, 71
lignosulphonate, 71, 75
Lockhart, T.P., 71, 73
lost circulation, 86
low tension polymer flood, 281

M

Martell, A.E., 207
McManus, D., 207
meteoric water, 122
Methuen, C.M., 99
Miano, F., 71
mineralogy, 16
Minton, R.C., 1
mixed metal hydroxide, 88, 95
mixed metal salts, 235
mobility ratio, 282
Monte Carlo Simulation, 21
mud, oil based, 13
mud, polymer, 17

O

oilfield reservoir souring, 179
oil price, 2
oil reserves, 4
oil reservoir, wettability, 251
operating costs, 5

P

PARCOM, 29
partially hydrolysed
    polyacrylamide (PHPA), 13, 57, 88
partnering, 8
Pelham, S.E., 99
phase separation, 56
Piro, G., 220
Plank, J., 97
plastic viscosity (PV), 86
polyacrylamide (PA), 88
polyalpha-olefin (PAO), 31
polyanionic cellulose (PAC), 88
polycarboxylic acids, 164
polyglycol, 57
polymer complexes, 290
polymer flood, 287
polymer gradient, 282
polymeric phosphino-carboxylates,
    149
polyvinyl alcohol (PVA), 298
pore pressure, 15
potassium, 57
produced water discharge, 311
Prudhoe Bay, 8
Przybylinski, J., 164
Puckett, D.A., 272

Q

Quad fields, 122
quaternary amine, 63

R

radioactive tracer, 40, 251
radioactivity, 321
Ramstad, K., 126
Reid, P.I., 13
reservoir souring, prediction, 180
reservoir wettability, 252
rheometric thickening, 99

S

Salters, G., 164
Sawdon, C.A., 28
scale formation, 116
scale inhibitor,
    adsorption isotherms, 126, 135
    divalent cation tolerance, 164
    effect of brine, 165
    mechanism, 165
    precipitation types, 127
    predictions, 127, 143
    retention, 127
    returns, 127, 135